BORN CANNIBAL

BORN CANNIBAL

Evolution and the Paradox of Man

James Miles

Foreword by
George C. Williams

IconoKlastic

Published by IconoKlastic Books
23 Millennium Place
204 Cambridge Heath Road
London E2 9NL
www.borncannibal.com
email@borncannibal.com

© James Miles, 2003
Foreword © George C. Williams, 2003

All rights reserved. No part of this publication may be reproduced, stored in a retrieval system, or transmitted, in any form or by any means, without the prior permission of the publisher. Within the UK, exceptions are allowed in respect of any fair dealing for the purpose of research or private study, or criticism or review, as permitted under the Copyright, Designs and Patents Act, 1988.

The moral rights of the author have been asserted

A CIP Record for this book is available from
the British Cataloguing in Publication Data Office

ISBN 0-9545552-0-1

Designed by Mitchell Davies, London
Cover illustration by Alan Aldridge, Los Angeles
Typeset in 9.75/12.75pt GaramondITC

Printed in Great Britain by
Butler & Tanner, Frome, Somerset

Contents

Acknowledgements		vi
Foreword by George C. Williams		ix
Introduction		xi
1	"He lets us down"	1
2	A complete disregard	17
3	Natura non facit saltum	37
4	A world without compassion	59
5	Casualties of war	82
6	Blessed of Nature	96
7	Magnificent exceptions?	111
8	The "delusion" of Free Will	137
	Endnote: Damned Man Walking: Christianity Meets Darwin's Wager	158
9	Lighting up hell-fires in Christendom	163
10	"Take me to your Prophet"; and other essays	176
11	Huxley's Paradox	196
Afterword		210
Letter: G. C. Williams to the author (03 December 1998)		212
Bibliography		213
Index		223

In memory of Darwin's Bulldog, 1825-1895

*I, too, wish to God there were
more automata in the world like you*

Acknowledgements

When the founding father of "selfish gene" theory recommended *Born Cannibal* to one of New York's largest publishing houses it was declined on the grounds that it would provoke considerable "anger" if published in America. Given that so few people have cared that the reading public learn the truth about the human genetic code, those who have helped get *Born Cannibal* off the ground deserve enormous thanks, and not just from me.

My love and gratitude to the family who have helped keep me sane over the last five years. To my mother and father, to Nicola and Antony Bream, and Shebnem Zorlu. Especial thanks to my brother, Dr. Chris Miles, who first got me to read *The Selfish Gene* even as he was sparking my interest in philosophy. Many friends and colleagues have been supportive of this venture, but I must single out two in particular. Rod Mackenzie, for his insights into the publishing world and his endless good humour (even when it was directed at *Born Cannibal*), and Dr. Yorick Rahman, friend and general sounding-board.

Thanks to my agent, David Grossman of the David Grossman Literary Agency. It is not easy for a London literary agent to be savaged by

Acknowledegments

indignant UK publishers who have taken offence at a book's central message and think that those with a background in philosophy have no right to be producing revolutionary pop science. Thanks, too, to Kirsty Gordon at the same agency. A big thank you to Mitch Davies for giving me a wonderfully designed finished product, and for suggesting using the innocence of the newborn for the cover. Thanks also to graphic supremo Alan Aldridge for the cover image itself, and for his haunting take on the Ouroboros archetype. This clever reworking of a millennia-old representation of the completion of nature so perfectly summarises the theme of *Born Cannibal*. Thanks to Razi Mireskandari at Simons Muirhead & Burton; not only one of London's leading defamation lawyers, but Partner at a practice with a deep commitment to freedom of speech and human rights. Not all who work in the legal profession merit the opprobrium Darwin, and Chapter 8, will heap.

Few academics emerge from the last 25 years of human evolutionary theorising able to hold their heads high, but I have found two who can. I would like to extend my thanks to the Editor of *Philosophy*, Anthony O'Hear. There are still philosophers out there who care more about the quality of the arguments than the non-existent reputations of those who are making them.

And, finally, one HUGE debt of gratitude to George C. Williams, Professor Emeritus in the Department of Ecology and Evolution at the State University of New York at Stony Brook. His expert advice and constructive suggestions on the text made this a better book than it could ever otherwise have been. The agreement by this doyen of evolutionary biology to provide the Foreword to a work by an unknown reflects both Professor Williams' remarkable lack of conceit and the importance he places on the advancement of knowledge. Dan Dennett says that it was George Williams who showed him how hard it is to be a good evolutionary thinker, and that Williams plays this game "better than anybody else in the world". A neuroscientist once likened working in modern evolutionary theory to wading through a cesspool. After five years of wading, sadly I cannot disagree, which is why it was all the more touching to come across someone who is worthy of keeping Darwin's name alive.

Foreword

The cavalries charged
The Indians died
Oh the country was young
With God on its side.

lines from a popular song by **Bob Dylan***

This book is cleverly written, biologically informative, and logically compelling. I expect it to become a classic work on the moral unacceptability of evolution by natural selection, a theory that biologists almost universally accept. This means, as Thomas Huxley argued more than a century ago, and Richard Dawkins more recently stated in modern terms, that we must rebel against the tyranny of the selfish replicators. This book will, I hope, effectively champion the Huxley-Dawkins condemnation of nature.

In 1993 I published a related essay entitled "Mother Nature is a Wicked Old Witch". A better title would have been "The Creator is Wicked but Fortunately Stupid" as implied by my 1996 book "Plan and Purpose in Nature". Human hope lies in that divine stupidity. The creator designed us for life in the stone-age environment, where benign relations with friends and especially relatives could be favorably selected because they

---

*Lyrics from "With God on Our Side" by Bob Dylan. Copyright © 1963 by Warner Bros. Inc. Copyright renewed 1991 by Special Rider Music. All rights reserved. International copyright secured. Reprinted by permission.

Foreword

could favor the survival of our selfish genes. Locally adapted tribal leaders could induce (or force) their associates to favor those selfish genes. The advent of agriculture and villages with hundreds of people for many generations made our environments grossly abnormal and not that to which we were adapted by natural selection. We could react to people in pulpits or on thrones as if they were talking to us individually, like friends or relatives or tribal leaders. Later, in a grossly abnormal environment, King Henry V (according to Shakespeare) could effectively address his troops as "we band of brothers" and inflict his own adaptive lie on thousands of enforced followers. Today we can be so influenced by preachers on television.

Obviously I am an antitheist and much opposed to atheists like my personal friend and theological enemy David Sloan Wilson. He has labeled himself an atheist, for instance in the New York Times of 24 December 2002, and I think his atheism explains why he fails to condemn the creator that designed the basic Darwinian process. Instead he champions a locally benign form of group selection that enhances the collective fitness of large groups of organisms. I think this kind of selection unimportant, but suppose I am wrong. Would God be less evil by favoring selfish groups rather than selfish individuals?

I doubt that James Miles really agrees with everything in these previous paragraphs or in my earlier publications. Despite my general praise I, of course, do not agree with every statement in his book, and there are topics that I wish he had discussed in more detail, like why nearly all multicellular animals reproduce with the complex and genetically wasteful sexual process, many in no other way. I think his praise for a few biologists, like George Williams and John Maynard Smith, a bit excessive, and likewise his condemnation of others, such as Martin Daly, Margo Wilson, and Helena Cronin, but these are minor disagreements. I have major agreements with his many and superb arguments against the God-is-good fallacy.

George C. Williams, Suffolk County, New York

Introduction

The dominant male, Humphrey, held a struggling infant about 1.5 years old, which I did not recognise. Its nose was bleeding, as though from a blow, and Humphrey, holding the infant's legs, intermittently beat its head against a branch. After 3 minutes, he began to eat flesh from the thighs of the infant, which then stopped struggling and calling.

David Bygott – "Cannibalism among wild chimpanzees"

For almost three decades now it has been the unexamined consensus in popular biology that our species managed to out-evolve the four billion-year pattern of the rest of nature. The thought that humans carry nearly the same genetic code as chimpanzees, for example, was, for many, too frightening to even consider …

When David Bygott published his paper in the journal *Nature* in 1972, little could he have known that he was at the forefront of a paradigm shift in evolutionary biology. Bygott concluded that there was insufficient data at that time to indicate whether the infant-eating behaviour he had witnessed in a group of chimpanzees was evolutionarily adaptive or just aberrant. Within a few years, however, the new worldview of gene-centred evolutionary theory (commonly known as "selfish gene" theory) that swept through biology would explain that such behaviour was indeed adaptive.

New field studies on primates specifically looking for cannibalism and

Introduction

infanticide catalogued shocking levels and distributions of such behaviour. Sarah Hrdy found infanticide to be the single greatest source of the up to 83 percent infant mortality rate found amongst the langur monkeys she studied in India. While cannibalism subsequent to infanticide is rare in primates, primatologist Mariko Hiraiwa-Hasegawa notes that the frequency of cannibalism in chimpanzees is now known to be "exceptionally high". As Hrdy dryly put it back in 1977: "we are discovering that the gentle souls we claim as our near relatives in the animal world are by and large an extraordinarily murderous lot".

Selfish gene theory has no problem with such behaviour. Selfish gene theory actually predicts such behaviour. As George Williams, the founding father of gene-centred evolutionary theory, puts it, the code of nature is a code of "gross immorality". Yet this understanding threw up a terrible dilemma for those who wished to use genetics to explain the human animal. If the code of nature was a code of gross immorality, and humans were the products of natural selection, how was morality even possible? Chimpanzees live in "a world without compassion", says primatologist Frans de Waal. So where did compassion come from?

The question was not new. Back in the nineteenth century Alfred Russel Wallace, the co-discoverer of the theory of evolution by natural selection, called morality the "inverse problem" of his and Darwin's work: how "to account for facts which, according to the theory of natural selection, ought not to happen". Darwin's Bulldog, Thomas Huxley, described nature as morally indifferent, and morality as the apparent paradox of natural selection. And Darwin, too, accepted that his theory of selection at the level of the individual could not explain morality. The question was to be asked again in the late twentieth century, but this time with much more urgency, because biologists now knew that nature was not just morally indifferent, but downright malevolent. The implication was obvious and, for many, terrifying. If chimpanzees, our closest living relatives, carry the genetic code for cannibalism, surely humans must too? **Surely natural selection predicted that all humans must carry the genetic code for cannibalism?**

So startling was this implication that selfish gene theory splintered into two factions over the issue in the mid 1970s. One faction, the larger and more vocal faction, insisted that humans *must* have out-evolved the code of the rest of nature, must have out-evolved the genetic code that

Introduction

nature had been using for four billion years. The other faction, the smaller and quieter faction, realised that nature simply could not reprogramme in this way. But the problem for those who wanted to argue that human nature had out-evolved the code of the rest of the natural world was that the vocal faction included no scientist who had ever made even a small contribution to developing selfish gene theory. Every single scientist who had actually helped establish selfish gene theory - from Williams to, amongst others, John Maynard Smith, Bill Hamilton and Richard Dawkins (who would coin the term "the selfish gene") - would end up joining Darwin, Wallace and Huxley in concluding that orthodox natural selection *could not explain human morality*.

Selfish gene theory predicts that all humans carry the genetic code for cannibalism. From Plato to Pope Pius IX, from presidents to prime ministers; all have carried the genetic code for cannibalism. Jesus Christ, who according to the Nicene Creed came down from heaven "and became human", must have carried the genetic code for cannibalism. You carry the genetic code for cannibalism, and so do your children. Selfish gene theory holds that morality *cannot* be explained genetically[1].

This is a book about the one ape that managed to escape from its biology, and consequently develop morality and civilisation. But this is also a book about evolutionary orthodoxy and evolutionary heresy, and about why, for a quarter of a century, the orthodox scientific establishment remained silent about that heresy. This is the story about how fear, ego and ideology blinded both scientists and philosophers. This is the tale of how Darwin was betrayed by his friends, as well as by his enemies ...

---

[1] This book is an exposition of Huxley's Paradox, which I first wrote about in the journal *Philosophy* in October 1998. However, it radically expands on the theme to examine a number of areas where scientists (and philosophers) have been refusing to treat man as just another product of evolution.

1

"He lets us down"

But when Darwin turns to his other solution, he lets us down.
Helena Cronin – The Ant and the Peacock

Darwin lets us down. This is the conclusion of Helena Cronin, one of the most prominent of the "evolutionary psychologists", the latest in a long line of people keen to proclaim that man is largely the product of his biology. Darwin's failure, for Dr. Cronin, was to suggest in his 1871 work *The Descent of Man* that culture, and not biology, was the key to human morality. Darwin had dared to say what most philosophers from Plato and Aristotle onwards had been saying: that in human beings environment, and not biology, is ultimately what counts. Darwin's crime, as you shall learn, was to believe that *Homo sapiens* is just another ape. For the last twenty-five years vocal human Darwinism, under such guises as evolutionary psychology, has been desperately trying to deny that man is just another product of natural selection.

Darwin's theory of evolution by natural selection destroys so many of

"He lets us down"

our comforting illusions, so many of our prejudices and so much of our arrogance. Darwin once confided to a fellow naturalist that the theory of natural selection "is like confessing a murder". In this book you shall learn more about the human implications of natural selection than you ever wanted to know, and it will sometimes make uncomfortable reading. This book strips away the myth to confront the reality of evolution. Darwin did not let us down. Darwin's understanding is repeated today, in far more strident terms, by the leading "selfish gene" biologists. And for the simple reason that nature is even nastier and further from morality than even Darwin could have realised in the nineteenth century.

Evolution by natural selection
"My initial interest in Darwinian theory was roused by philosophers' criticisms – not because I thought that they were right but because I was convinced that they must be seriously wrong", wrote Helena Cronin in the introduction to *The Ant and the Peacock*. She had always been convinced, Cronin tells us, of the fixed structure of human nature. She would dispute, for example, Aristotle's belief that we are born without any character at all, and that whether we acquire a good or bad character depends on the kind of upbringing we get[2]. Aristotle's view, as given in his *Nicomachean Ethics*, was his conclusion after decades of studying the divergent characters of the Greeks found in the multitude of Greek city-states. Aristotle's point on human malleability has been remade by many philosophers across the millennia. As Nietzsche was to write over two thousand years later in *Thus Spoke Zarathustra*: "Much that seemed good to one people seemed shame and disgrace to another: thus I found. I found much that was called evil in one place was in another decked with purple honours". While their views on the nature of goodness were markedly different, Aristotle and Nietzsche were as one in noting the incredibly powerful moulding effects of culture.

But in believing that humans are born with no character at all Aristotle was indeed wrong. Cronin is right on this point. In believing that humans are born with no character at all Aristotle is refusing to treat us as animals. As we shall see in a later chapter, all animals are born with a

---

[2] This is the understanding of Aristotle given by the noted Aristotelian scholar J. O. Urmson. See chapter 2 of his *Aristotle's Ethics*.

universal nature; we are animals and so humans, too, must be born with a universal nature. But Aristotle was wrong for the right reasons. He thought we are born with no character at all because his own eyes showed him the immense malleability of humans. The error Dr. Cronin makes is to assume that the universal nature we are born with must equate to a universal human nature. Her mistake is to assume - against all the Darwinian evidence - that biological nature and human nature are one and the same, and that culture has no substantial effect on biology. Darwin did *not* make this mistake, and this is why Cronin tells us that "he lets us down". For Darwin something immense must be happening to us after we are born, because we are born selfish apes - a biological fact Dr. Cronin is effectively denying. Cronin, and the traditions she represents, appears no more willing to treat us as just another animal than Aristotle was.

Evolution, the understanding that species change with time to form entirely new species because of the inheritance of different characteristics, was not new when Charles Darwin produced his work. "Transmutation", as it was then called, was already in the air when Darwin published *Origin of Species*. Charles' grandfather, Erasmus Darwin, was among those who had produced early theories of "perpetual transformations" within nature. The Frenchman Jean-Baptiste Lamarck had proposed an even more influential, and alternative, theory of evolution that Darwin himself toyed with for many years before ultimately rejecting. And Robert Chambers' best-selling *Vestiges of the Natural History of Creation* had already prepared Victorian society for the fundamental shift in world-views, as the old Biblical static ordering of separately created species would be swept away to be replaced by a transmutationary landscape. But what Charles Darwin and his contemporary Alfred Russel Wallace were to discover was one particular and very convincing explanation for evolution, that of evolution by "natural selection".

What we now know is that natural selection applies to entities with the following characteristics: multiplication, variation and heredity. What this means is that natural selection works on entities that can make copies of themselves. Variation means that copying, however, will never be perfect, and it is random copying errors that are essential if the process of evolution is to occur. Once the error or "mutation" has occurred, heredity will ensure that a mutation reappears in all future

"He lets us down"

generations. Over time evolution will tend to favour higher multiplication, longevity (the ability to survive long enough so as to be able to reproduce) and heredity. Natural selection is the process that determines that traits which are conducive to superior replication (Darwinian "fitness") will, of mathematical necessity, tend to become increasingly common in a population over time.

Adaptation was key to Darwin's explanation of evolution by natural selection. Adaptation, or perfection of design, is explained as the gradual adjustment in form and behaviour as selection saves what is useful, and discards what is less useful, and slowly improves designs to fit an environment. Natural selection can this way often produce extremely precise - though never perfect - contrivances. Gradual saving of slight improvements, will, over very many generations, allow a blind, physical process of accumulating small beneficial mutations to create the wonders of nature that the late eighteenth-century theologian William Paley had seen as proof of God's intervention; the eye or the hand. Paley's assumption of a forward-looking Creator necessary to fashion the most perfect designs in nature - natural equivalents, said Paley, of such perfect human creations as the watch - falls away. The "blind watchmaker" of natural selection becomes the main mechanism for creating the wonderful diversity of plant and animal life we see around us.

Human beings are animals. Human beings are products of natural selection. Where vocal human Darwinism has been going wrong is in refusing to accept this understanding; in trying to argue that we evolved under different rules, products of a more benign creative process. Human beings do not act like any other species of ape. Our behaviour, as you shall learn, often seems to be diametrically opposed to chimpanzee behaviour. But instead of accepting that we evolved under the same pattern of natural selection as all other animals, and *then* looking for reasons for our unique behaviour, evolutionary psychologists and others suggest that mankind evolved under an entirely different pattern. They suggest that, 150 to 100,000 years ago on the savannahs of Africa, our ancestors broke from the immutable pattern that had existed in nature for almost four billion years. As you shall learn, they are so very wrong.

Civilized human behavior has about as much connection with natural

selection as does the behavior of a circus bear on a unicycle.
Mark Ridley & Richard Dawkins – "The Natural Selection of Altruism"

Human sociobiology

In many ways the groundwork for what was to come in the form of human sociobiology had been laid in the late 1950s by the Nobel Prize-winning linguist Noam Chomsky. Chomsky had shown that the capacity for language acquisition is an innate, inborn, capacity within the human animal; that we have a natural language "organ". Many cognitive theorists since then have tried to argue that the demonstrable genetic capacity for language acquisition is evolutionarily equivalent to the suggestion that our behaviours have a strong genetic component. The inevitable result was what would become known as "human sociobiology". Chomsky himself was to strongly refute such suggestions, but as the science writer John Horgan put it in *The End of Science*: "Edward Wilson and other scientists who attempt to explain human nature in genetic terms are all, in a sense, indebted to Chomsky".

In 1975 a Harvard entomologist named Edward O. Wilson, a recognised expert on insect behaviour, was to coin a new term. "Sociobiology" was defined as the scientific study of the biological basis of all forms of social behaviour in all kinds of organisms, including man. In his magnum opus *Sociobiology* he summarised the previous five decades of research into animal social behaviour by biologists all over the world. The book would have been little more than a useful reference tool for fellow biologists were it not for the one small chapter that dealt with man. Other animals could be explained in terms of their genetics; for Wilson man, too, was little more than his genes.

Wilson did not single-handedly invent the recent rush to reduce virtually every facet of human existence to genetic terms. But he gave the movement strong impetus with his suggestion that all manner of human behaviors – from schizophrenia to religion to criminality – have a strong genetic component.

Niles Eldredge – Reinventing Darwin

As we have already seen, philosophers have been making a grave mistake when they refuse to see man as another animal (albeit a very special animal). Richard Dawkins, the author of *The Selfish Gene*, was

"He lets us down"

quick to defend Wilson's programme by repeatedly pointing out that sociobiology was defined as a neutral field of study. Therefore, by definition, it could not be false, since it sought only to establish what elements (if any) of human behaviour had a biological component. "It is a field of study ... not a point of view", he wrote in his essay "Sociobiology: the New Storm in a Teacup". Yet even as he was writing this, Dawkins was to display some unease, not at the exercise, but at its application. "If I were to encapsulate in a phrase the stance, or commitment, of modern sociobiologists, I would say that they are Behavioural Darwinists. So, of course, are ethologists, and incidentally I therefore see no very great need for the word 'sociobiologist' to have been coined at all." Dawkins explained that in his own view it is "downright naïve" to look at the social life around us and try to interpret the actions of individuals directly in terms of survival value or gene preservation. "The operative word is 'directly'."

Wilson's own protestations that he was involved in a neutral field of study, and not in putting a particular point of view, were becoming a little hard to sustain. His claims in *Sociobiology*, made more explicit still in a number of newspaper articles and in his 1978 work *On Human Nature*, left few in any doubt that he had gone beyond study and had already formed his own concrete opinions. According to Wilson most things had a strong genetic component. Love, grief, nationalism, aggression and most other human emotions and behaviours. Even sexual preference. Wilson considered whether homosexuality in humans might be a Darwinian adaptation analogous to worker sterility in ants and the other "social insects" he had spent decades studying. Human homosexuals, some sociobiologists suggested, might not care about reproduction because they ensure survival of their genes by helping to raise relatives who carry many of the same genes.

Natural selection created our similarities, said Wilson, but it also created many of our differences. All humans carry slight variations in their genetic make-up, and (Wilson argued) this is largely responsible for the observable behavioural differences between humans. In *On Human Nature* Wilson argued that common decency and morality had evolved in our ancestors under Darwinian laws, a claim he first made explicit in his 1975 article "Human Decency is Animal". Or as the human sociobiologist John Gribbin has put it: "helping the sick and the weak ... emerged through the process of natural selection" (Gribbin and Gribbin

[1993]). Therefore while most people are born with good genes, some (by mutation, or inheritance from a bad gene pool) are born with bad genes. Sadly, concluded Wilson, we must abandon the left-liberal dogma that says otherwise. In the genetic lottery, some are born winners, and some are born losers.

This argument that humans had evolved decency was to be the keystone of all subsequent human sociobiology. For anyone determined to show that human behaviour is dominated by genetic considerations providing a rationale for all the most valued human characteristics becomes of overriding concern. Decency, morality, virtue, honour, patriotism; they all need a genetic basis. So, for Wilson, large-scale human self-sacrifice exists because human evolution has worked in such a way that the individuals within the species did better by evolving *unconditional disinterested* altruism. This is described by him as "hard-core" altruism, the result of a combination of factors including perhaps a genetic capacity for blind conformity. The human sociobiologist and Darwinian philosopher Michael Ruse tells us in *The Darwinian Paradigm* that human evolution has worked in such a way that "one will probably function most efficiently when one has no hope of return at all. … [T]he claim is that (literal) altruism and (biological) 'altruism' are connected. In particular, it is argued that, in the case of humans, in order to make us perform 'altruistically', because we do indeed (for good biological reasons) have selfish feelings, we have laid over us (literal) altruistic inclinations". The position of the modern evolutionist, notes Ruse, is that humans have an awareness of morality, because such an awareness is of biological worth. "Morality is a biological adaptation. … Perhaps we really ought to hate our neighbours, but we, poor fools, think otherwise!"

But many Darwinians remained to be convinced by the theorising of the early human sociobiologists. For example, as we shall see later, social insects have very special inheritance and reproduction systems that can be used to explain why they act the way they do. Inheritance and reproduction systems that we do not share. The suggested analogous model for human homosexuality suffers from serious theoretical weaknesses. In 1981 Dawkins and his zoological colleague at Oxford Mark Ridley coined the term "evangelistic" sociobiology to refer to the many sociobiological hypotheses then around that were claiming far more for evolution than biological orthodoxy allowed. "More recently,

"He lets us down"

Darwinism has suffered by the enthusiasm for swallowing other disciplines that has marked sociobiology's more evangelistic texts. Thus Wilson (1975) has written: 'In this macroscopic view the humanities and social sciences shrink to specialized branches of biology'." It would be a shame if such brash overestimation of the explanatory power of evolutionary biology were to blind social scientists to the importance of Darwinism for their subject, wrote Ridley and Dawkins.

Though most of his criticism of these "evangelistic" sociobiologists is restricted to less mainstream publications, Dawkins was to lash out at them, albeit rather briefly, in the second edition of his bestselling *The Selfish Gene*. Many of them have been angry at his refusal (for the reasons we shall investigate) to join their mission to persuade us all that genes are what really count. Buried in an endnote to *The Selfish Gene* he fumes briefly at "the criticism from doctrinaire sociobiologists jealously protective of the importance of genetic influence" in the human animal. It is indicative of the problems that beset modern human evolutionary biology that while his gene-worshiping critics who have betrayed Darwin's legacy rate a mere 20-word tongue-lashing, the late paleontologist Stephen Jay Gould, who has at least made a number of important contributions to evolutionary theory, has been savagely attacked by Dawkins at every opportunity for over twenty years. We shall investigate how this sorry state of affairs - a state of affairs that ultimately led to the betrayal of Darwin's legacy by all players in the Darwinian drama - came to be in chapter 5.

Others have been less willing than Dawkins to make only passing comment on bad Darwinism. Stephen Jay Gould is probably the most widely read evolutionary theorist in America. His numerous books and regular newspaper articles have helped bring the Darwinian message to many, many people. But most of his readers are probably unaware that he represented the less orthodox of the two major traditions within modern Darwinism. "Selfish gene-ery", the tradition led by George Williams and Richard Dawkins, the (supposed) tradition of all human sociobiologists, and the tradition to which I belong, is actually the mainstream tradition today. We shall consider the debate between Gould's tradition and the far more widely accepted selfish gene tradition in a later chapter, because the debate goes a long way towards explaining why so many errors have crept into *human* selfish gene theorising. Nevertheless we need not consider the debate itself for a few chapters because it is only incidental to the errors that exist within the one tradition of selfish gene-ery. Gould's contribution to exposing those errors has been praiseworthy (even if his motivation can be somewhat questioned). Gould was one of the first to forcefully note that early human sociobiologists were often telling what he termed "evolutionary Just So Stories", tales which were based on spurious analogies or unproven assumptions. As with Rudyard Kipling's original *Just So Stories*, basing fanciful speculation on unproven (and often unprovable) assumptions is not good science, and it may be little more than storytelling. Some within the tradition Gould represents (and notably the geneticist Richard Lewontin) found fault with their genetic models; others from a more philosophical background (such as the philosopher of science Philip Kitcher) pointed to logical and methodological errors.

Under the relentless hostility of Gould's tradition (which sometimes, it should be noted, resorted to less wholesome methods such as character assassination), human sociobiology was largely dead by the mid 1980s. The leading selfish gene theorists did little to even try to keep it alive; this had never been their worldview, after all, although no one even bothered to ask *why* they had not supported it. Few today (Wilson excepted) will even wear the badge "human sociobiologist". From promising to overturn the social sciences, Wilson was effectively gone little more than a decade later. Gone, but his dreams not forgotten.

Evolutionary psychology

As the evolutionary psychologist Steven Pinker puts it, "there is a universal design to [] the human mind". But this is no longer to say there is no moulding of human character. "People are flexible, not because the environment pounds or sculpts their minds into arbitrary shapes, but because their minds contain so many different modules, each with provision to learn in its own way".

Evolutionary psychology (or "son of sociobiology", as John Maynard Smith has called it) is claimed to be very different from the human sociobiology that it replaced in the mid 1980s. Evolutionary psychologists profess that social scientists misunderstand them; that, unlike both behavioural geneticists and many of the earlier human sociobiologists, they hold that the great majority of human differences are cultural, not genetic, in origin. As John Horgan put it in his 1995 article for the journal *Scientific American*: "[T]his view conforms to the

"He lets us down"

party line of evolutionary psychology, which holds that with the important exception of sex, all humans are born with essentially the same psychological endowment".

Human sociobiology had often given the answer that the differences between people's behaviour are the results of the slightly different genetic codes they are born with. Evolutionary psychologists now professed to reject this argument. Horgan quotes from their literature: "'Evolutionary psychology is, in general, about universal features of the mind. ... Insofar as individual differences exist, the default assumption is that they are expressions of the same universal human nature as it encounters different environments'". In other words our similarities are genetic, but our differences are assumed to be cultural, and this was a considerable shift in the basic worldview of this son of sociobiology. The main reason that evolutionary psychologists distanced themselves from the theories of the human sociobiologists will be discussed in the next chapter when we begin to study selfish gene theory. For the moment it is enough to understand that they do. As John Horgan noted in 1995: "In fact, the surest way to annoy evolutionary psychologists is to lump them together with behavioral geneticists [*and human sociobiologists*], who tend to ascribe differences among individuals and even ethnic groups to genetic variation". So contrary to human sociobiology, evolutionary psychology stopped denying that natural selection created a species-wide behaviour pattern in man. Human similarities were seen as genetic, but human differences were accepted as being predominately cultural. Where individual behavioural differences occur they are supposed to be the result, not of different genetic codes, but of different genetic expression. The new idea was that genes create mental "organs" that will express themselves one way in one environment, and another way in another environment. Evolutionary psychology was therefore largely *confirming* what philosophers and social scientists already believed: that environment moulds us, that environment, and not biology, is responsible for our varied behaviours. Surely now Darwinism and the social sciences could be reconciled? And many social scientists did indeed flock to this new call, often becoming the most zealous advocates of the new understanding.

But there remained a fatal flaw to all evolutionary psychology. Central to evolutionary psychology is this idea of Wilson's that we evolved to be "the decent ape", though the justifications have often changed. Wilson asserted direct evolution of morality, but (for reasons we shall come to) even evolutionary psychologists found it difficult to accept this position. Instead they tried to find other arguments. In one of the earliest major works of this new discipline we were told that nature had not so much selected for morality as produced a benign evolutionary accident through environmental change. Excessively moral behaviour could be accounted for because we are "adaptation-executors" and not "fitness-maximizers" and as such, while unconditional altruism never evolved *per se*, it is the evolutionary consequence of 100,000-year-old selfish behaviour that in today's alien environment has the manifested expression of selflessness[3]. One of the earliest popularisers of the new learning, the science journalist Robert Wright, gives us his key to human altruism in *The Moral Animal*: "[i]n sum, the best guess about valor in wartime is that it is the product of mental organs that once served to maximize inclusive fitness and may no longer do so".

Yet just because human beings may often demonstrate certain similar behaviours, is this any reason to assume, as evolutionary psychologists tend to do, that they are genetic in origin? The Darwinian philosopher Daniel Dennett is keen to point out in his bestselling *Darwin's Dangerous Idea* that we cannot simply make this assumption. "[S]howing that a particular type of human behavior is ubiquitous or nearly ubiquitous in widely separated human cultures goes *no way at all* towards showing that there is a genetic predisposition for that particular behavior", he comments (emphasis his). And unfortunately the tendency to tell speculative "Just So Stories" did not end with human sociobiology. Evolutionary psychology must make good use of them too. Dennett singles out Donald Symons' essay "On the Use and Misuse of Darwinism in the Study of Human Behavior" as a "bracing exception" to this trend; as one willing to do battle with the greedy claims of some human sociobiologists and evolutionary psychologists. Unfortunately Symons, as an evolutionary psychologist, is less willing to apply such criticism to himself. In the self-same essay where he is supposed to be bravely combating the greedy claims of others, Symons makes the following suggestion: "In fact, since the adaptations that

---

[3] J. Tooby & L. Cosmides, "The Psychological Foundations of Culture", also D. Symons, "On the Use and Misuse of Darwinism in the Study of Human Behavior". Both in *The Adapted Mind* (eds: Barkow & ors.).

"He lets us down"

underpin human behavior were designed by selection to function in specific environments, there is a principled Darwinian argument for assuming that behavior in evolutionarily novel environments will often be *mal*adaptive." Or, to translate, the adaptation that produced behaviour A 100,000 years ago might actually manifest itself as behaviours X, Y or even Z, once our ancestors left the African plains. In other words, we are back to the unfalsifiable loose speculation that characterised human evolutionary theorising in the 1970s.

So one problem with evolutionary psychology is that it trades in the same unfalsifiable "Just So Stories" as human sociobiology. For every hypothesis proposed by evolutionary psychologists there are alternative explanations. For every example cited to support a pet theory there are other examples that provide conflicting conclusions. As George Williams commented to John Horgan: "I keep thinking of counterexamples". But evolutionary psychology is not satisfied with unfalsifiable stories. As we shall see in the following chapters, evolutionary psychology often trades in theories that are at best implausible and often downright anti-Darwinian.

A model methodology?

Humans share almost 99% of their DNA with chimpanzees[4]. Yet one feature of sociobiological theorising (father and son) seems to be the constant need to treat us as something other than what we are - apes. Sociobiologists search for models in nature that will allow us to draw parallels with our own species. But their models not only tend to be deeply flawed, they tend to abandon all hope of finding natural examples within our own biological class (the mammals).

One wonderful example comes from the British science journalist and writer team of John and Mary Gribbin (who began supporting human sociobiology in its earliest days). On the back cover of the paperback edition of the Gribbins' 1993 work *Being Human*[5], Edward Wilson writes: "The Gribbins explain the basis and paradox of sociobiology in

---

[4] Calculated using base substitutions, i.e. taking account of sections of DNA found on the genomes of both chimp and human. Please see the footnote in Chapter 3 that addresses reservations with this method.

[5] *Being Human* was a re-issue (with the addition of one new chapter) of their earlier book *The One Per Cent Advantage: The Sociobiology of Being Human*, published in 1988.

language that can be understood by all, making it clear why the subject is provocative and unpalatable to many even as it tries to provide a unified view of human behaviour". The Gribbins tell us that we must search for natural "models" on which to base human sociobiology. The nearest model that they can find is not within the apes. It is not within the primates. It is not even within the mammals. It is, we are told, the birds. Birds have evolved true altruism too: "[t]he pattern of behaviour that makes for helping at the nest of your siblings is almost exactly the same as the pattern of behaviour that makes for helping at the nest of any member of your species. ... [H]elping becomes a common activity ... as genes for helping spread through the gene pool". Humans share no more genes with the birds than we do with lizards or snakes, so cherry picking a favoured model so far removed from our biological line is a little dangerous. While it is not automatically a non-starter to try to cross the biological family, or order, or even, in this case, class barrier in this way, it should give readers pause for thought and prod them into being a little more critical.

But the cherry picking is not even valid. In the third chapter we shall return to this bird model to consider the single greatest flaw with it, but for the moment we shall just consider its lesser problems. The Gribbins' analogous reasoning is a little strange. Natural world sacrifice for kin (or rather, the genes they carry) becomes true "niceness", we are told, because "some individual helpers, either through confusion, inability to recognize their kin, or even as the result of a very slight mutation, will help more widely ... such a tendency then begins to act 'for the good of the species', although it has its origins firmly in the selfishness of genes". Unfortunately this model, chosen specifically as an important "analogy" for human sociobiology, has little relevance: "[a] long memory and a capacity for individual recognition are well developed in man", writes Dawkins in *The Selfish Gene*. (As mentioned, the reason that we cannot invoke their third suggestion, "a very slight mutation", will be dealt with later.)

Another model with far more scientific accuracy, but even less validity, is the science writer Matt Ridley's suggestion in *The Origins of Virtue* (and he is not to be confused with *Mark* Ridley, the noted Oxford zoologist quoted earlier). The Gribbins sought to cross the class barrier. In his book Ridley goes one step higher and seeks to change phylum by linking us with the invertebrates. In the book he explained that group

"He lets us down"

cohesion comes naturally to us, that we are genetically programmed for co-operation from a foundation in reciprocity (or "you scratch my back, I'll scratch yours"). Now if he had left it at this he would not have been a heretic. If he had argued that our co-operative instincts are those shared with other apes (that is, small group reciprocity, and sacrifice for close kin) he would have been supported by orthodox evolutionary theory. But, as with all sociobiologists, he tries to go further. He tries to claim that human large group cohesion is somehow adaptive. And he uses the model of the ant to make his argument.

He begins by quoting two passages from others' works drawing attention to the common observance of ants working for their group, and to a long tradition of paralleling this with observed human behaviour. He then goes on: "[w]hat these two descriptions have in common is not just an instinctive comparison between the societies of social insects and human beings but a recognition that somehow the ants and the bees are better than us at doing something we strive towards". Their societies are more harmonious, Ridley tell us, "more directed towards the common, or greater, good". Putting aside the "instinctive comparison" for the moment, we should really be asking ourselves why we should expect humans to be striving towards "the common, or greater, good" in the first place. Biologists today know that ants and bees behave the way they do because of their unique genetic inheritance and reproduction systems. But what is our rational?

Mammals inherit half their genetic material from each parent. In some species of social insect females develop from fertilised eggs – they are called "diploid" and, like mammals, have a double set of chromosomes, one set from each parent. In contrast the males develop from unfertilised eggs – they are called "haploid" and have only a single set of chromosomes to pass on. Commonly a worker class will be female, and will be sterile and so cannot reproduce. Because their father was haploid, the daughters are identical in half their genes (their father's), and statistically will share half their mother's genes, giving an overall genetic relationship of three-quarters. They are then more closely related to each other than they would be to any notional offspring they might have had (had they not been sterile). In normal diploid species siblings share half their genes; in haplodiploid species sisters can share three-quarters of their genes, permitting natural world kin sacrificial behaviour to be particularly potent. The upshot is that the sterile workers' genes actually benefit more from their "sacrificial" behaviour of caring for their sisters who potentially remain fertile and can become future queens.

Now the problem for those who try to suggest any sort of link or analogy between social insect "altruism" and human altruism is that in social insects all we are seeing is another form of kin sacrifice caused by particular genetic and/or sterility issues. This form is generally as predictable on the degree of relatedness / degree of sacrifice matrix as is mammalian sacrifice. But crucially they are *different* matrices. Social insects have their genetic inheritance and reproduction systems, and mammals have theirs[6].

We are not haplodiploid, nor do we have sterile castes, nor do we have an inbreed/outbreed cycle. Our genetic inheritance system, separated from a common ancestor with the social insects by *hundreds* of millions of years, therefore has no such necessary implication. We would not expect chimpanzees to be striving towards this "common, or greater, good", yet we share their genetic model, their reproductive system, their breeding pattern. So why should we even *expect* our genes to manifest themselves differently from chimps? According to Ridley reciprocity (and not, as in ants, kin relationship factors) "overcomes" selfishness. As he commented on BBC Radio 4's *Start The Week* in October 1996: "… because individuals are inherently fairly selfish. So you've got to overcome that, and the way we have overcome it … is through reciprocity". Yet biological reciprocity cannot overcome selfishness because biological reciprocity *is* selfishness. To argue that genetics can take us beyond what is known as the "technical" altruism found in nature and lead us to "true" altruism, or to altruism beyond that which is observed in other primates, is simply untrue, as we shall come to discuss.

Returning to the "instinctive comparison", we get an inkling into why the sociobiologists cannot help but try to cut all ties with the mammals. Observation undeniably separates us from other vertebrate behaviour; man might be *from* the brutes but "he is assuredly not *of* them" as Darwin's Bulldog, Thomas Huxley, put it over one hundred years ago.

---

[6] George Williams pointed out to me in 1998 that termites are not haplodiploid, but achieve their powerful form of kin sacrifice largely through the evolutionary device of worker sterility (and a cycle of inbreeding and outbreeding).

"He lets us down"

But for a sociobiologist the answer must lie in genes. And in providing their suggestions they leave the world of hard Darwinism and enter a world of ever more wondrous "Just So Stories" and ever more fanciful models. Ridley tells us in *The Origins of Virtue* that "We are more like ants or termites who live as slaves to their societies", or, as he commented on *Start The Week*: "We are in a sense the ants of the ape family".

Ridley is not actually arguing that we are very tall ants, but he is looking for *biological* explanations (which he finds in natural world reciprocity, and which we shall deal with in the next two chapters) for why we exhibit their degree of co-operation. He realises that he cannot use haplodiploidy kin selection, or the sacrifice one organism makes for close relatives, but consequently his "analogy" is therefore meaningless. It is actually potentially worse than meaningless because it comes close to giving the dangerous impression to casual readers that human and ant genetics imply the same expected behaviours. Such a suggestion would be utterly wrong, and Ridley does take some pains to explain to his readers that this is not what he means. Yet since there is no genetic analogy (and no sterility correlate), why run the risk of choosing such a potentially misleading model? At least the Gribbins' rather misguided bird model shared the same basic genetic and reproduction patterns as man.

But why do the sociobiologists and evolutionary psychologists feel they have to look for models outside the other mammals? And what happens when we begin to regard ourselves merely as apes? Just what is so wrong with this core sociobiological hypothesis that we evolved morality and virtue? In the next chapter we will begin to get a very *different* explanation of human morality when we look at the work of both the great nineteenth-century Darwinians and today's leading selfish gene Darwinians.

2

A complete disregard

But natural selection ... implies concurrently a complete disregard for any values, either of individuals or of groups, which do not serve competitive breeding. This being so, the animal in our nature cannot be regarded as a fit custodian for the values of humanity.

W. D. Hamilton – "Selection of selfish and altruistic behaviour in some extreme models"

In 1859 in *On the Origin of Species* Darwin wrote that: "natural selection can act only through and for the good of each being. ... Natural selection ... will adapt the structure of each individual for the benefit of the community; [but only] if each in consequence profits by the selected change." Darwin was an "individual-selectionist"; he believed that nature could select only for the benefit of the individual. He here explicitly disavows "group selection", or the idea that nature will select for the benefit of the group, rather than its component parts. Any benefit to a group can only be as a by-product of what is good for its individuals.

A complete disregard

In 1966 George Williams launched the "gene-selectionist" (later to become known as the "selfish gene") revolution, where apparent individual selection was reinterpreted as, not what is good for an individual, but as what is good for its genes. Since the fate of an individual and the fate of its genes are very closely (but not perfectly) linked, individual selection is often for practical purposes synonymous with gene selection.

As Darwin himself realised, given that natural selection can act only for the good of each individual, altruism would appear to present problems for evolutionary theory. However, in the orthodox selfish gene tradition represented by Williams, Richard Dawkins and John Maynard Smith, natural world altruism has been explained through the two main mechanisms known as kin selection and reciprocal altruism.

Kin selection explains the evolution of altruistic characteristics towards close relatives because a gene for altruism can spread because it enhances its own replication through its survival effects on the relatives. The understanding that animals sacrifice for immediate kin was of course central to Darwin's theory, with fitness defined in terms of simple biological success. But in 1964 kin selective altruism became one of the building blocks of modern selfish gene theory after the publication of the late Bill Hamilton's work. Hamilton suggested that animals seek not to maximise their own fitness, but rather their own "inclusive fitness". The understanding is that genes pass into future generations not only through direct offspring, but also through relations. In normal diploid organisms, a parent shares a genetic relationship of one half with its child. But the parent also shares such a relationship with its siblings (because of inheritance from *their* shared parents). It will statistically share a quarter relationship with the sibling's offspring, and so on. Relations, and not just offspring, carry an individual's genes into subsequent generations, and so it makes sense (from the "point of view" of an animal's genes) for a degree of sacrifice for near-relatives. Hamilton took these relationships (which had earlier been noted by the geneticist J.B.S. Haldane) and provided a rigorous mathematical formulation of the altruism that could (and could not) be expected under such genetic inheritance. Kin-directed altruistic behaviour can be maintained under the selection process provided the cost of that behaviour to the altruist (in terms of reduced personal fitness) is less than the benefit of the behaviour to kin (in terms of inclusive fitness) multiplied by the coefficient of relatedness. Hamilton further extended the work to encompass haplodiploid inheritance, and provided a mathematical rationale for social insect eusociality in terms of the even closer three-quarter genetic relationship between sisters.

Reciprocal altruism is the exchanging of altruistic favours such that each benefits more from co-operating than it would from not co-operating (or, "you scratch my back, I'll scratch yours"). Darwin had noted that animals indulge in co-operation and altruistic acts towards those that may not be closely related where the acts seemed to be for the mutual benefit of each party. In 1971 Robert Trivers coined the term "reciprocal altruism" to describe the trading of favours within the natural world. Favours can be returned immediately, or at a later time provided the animals can remember interactions and recognise parties (as hypothesised by Darwin, and now termed "delayed reciprocal altruism"). A mutation that first promotes such behaviour can be positively selected.

Examples of reciprocal altruism in the natural world are numerous, and include the mutual grooming of primates, and even the oft-cited actions of the tiny cleaner fish. Various species of cleaner fish occupy specific locations, and offer services to larger fish by cleaning them of parasites and unwanted particles. The cleaner fish thereby obtains a reliable food supply from the visitor, as well as a degree of protection from the larger fish. The cleaned fish benefit from having dead tissue and such removed. The selfish gene insight of reciprocal altruism was then to be greatly developed and given a more mathematical basis through the work of John Maynard Smith, who was able to formulate individual strategies of co-operation and confrontation (see later).

Both kin selection and reciprocal altruism are sometimes referred to as "technical" altruism, to make the point that such altruism is ultimately self-serving. All conscious "altruism" in the natural world reduces to selfish genetic prudence.

The nineteenth century - Darwin, Wallace and Huxley

Nevertheless, in dealing with man's evolution there was one point (a point incidentally noted by Wallace as inexplicable through selection) where Darwin for once did quaver in his commitment to individual selection. This was over the evolution of the human moral sense.

Michael Ruse – The Darwinian Paradigm

A complete disregard

Charles Darwin

Darwin had been an avowed individual-selectionist for over three decades when he wrote *The Descent of Man* and seemed to begin to speak openly in terms of group selection. "[U]ncharacteristically, he seems quite explicitly to be offering a higher-level solution", writes Helena Cronin in *The Ant and the Peacock*. "Moving on to humans, however, Darwin does recognise that here there may be real self-sacrifice; moral considerations are likely to clash with our selfish interests, even overriding our bid for self-preservation".

Yet Darwin never abandoned his commitment to individual selection; he just realised that the amorality of natural selection could not explain human morality. Although nature had given us certain traits which could be built upon, or twisted, something else had to be doing the building and the twisting. Darwin realised that something *more* was needed. How could truly virtuous human self-sacrifice, where, said Darwin, the bravest men "freely risked their lives for others", come about? Darwin had two solutions. The first was his "low motive" of commonplace natural world reciprocal altruism. "But there is another and much more powerful stimulus to the development of the social virtues, namely, the praise and the blame of our fellow-men. ... Ultimately a highly complex sentiment, having its first origin in the social instincts, largely guided by the approbation of our fellow-men, ruled by reason, self-interest, and in the later times by deep religious feelings, confirmed by instruction and habit, all combined, constitute our moral sense or conscience", wrote Darwin. "A belief constantly inculcated during the early years of life, whilst the brain is impressible, appears to acquire almost the nature of an instinct."

"[W]hen Darwin turns to his other solution, he lets us down", Cronin writes. And yet Darwin was not abandoning his commitment to individual selection. As we shall see, his answer is effectively the same as Williams, Maynard Smith and Dawkins will all give. Darwin (unlike the evolutionary psychologists) is not arguing that man is born coded for decent and virtuous behaviour. The answer for Darwin was that culture took just another biologically selfish ape (the "first origin") but then twisted it into an altruistic human, through positive and negative peer-pressure, myth, instruction and habit[7].

---

[7] This is undoubtedly an oversimplification of *Descent of Man*. As Richards (1987) points

A complete disregard

Alfred Russel Wallace
Alfred Russel Wallace: "[T]he ever-vigilant defender of natural selection, the ultra-adaptationist, the most Darwinian of Darwinians. And yet, when it came to humans, particularly to our moral sense ... Well, here are Wallace's own words: 'It will ... probably excite some surprise among my readers to find that I do not consider that all nature can be explained on the principles of which I am so ardent an advocate'", writes Helena Cronin in *The Ant and the Peacock*.

Wallace is today the great footnote in Darwinian history. This surveyor-turned-naturalist independently developed a remarkably similar theory of evolution by natural selection at a time when Darwin was still conducting his investigations. Darwin had sat on his own theory for twenty years, continually refining it, and was only finally pushed to publish when Wallace, suffering from malaria in South East Asia, sent him his own essay. Wallace and Darwin's unpublished works were thus presented together at the Linnean Society in July 1858. However, because of the more popular appeal of Darwin's subsequent *Origin of Species*, and Thomas Huxley's efforts to sell Darwin's message, it was assured that it would forever be Darwin's greater contribution that is remembered. While Darwin was being fêted at home, Wallace continued working in the field, later even becoming a founding father of biogeography, the study of the geographical distribution of living things. It was Wallace who first recognised that the flora and fauna of South East Asia are split between Asian and Australian types. It was Wallace who would give his name to Wallace's Line, the sharp zoogeographical divide that cleaves the Indonesian islands into two realms, a recognition that would answer contemporary questions in speciation, and even questions in plate tectonics a century later. Although greatly honoured in his lifetime, Wallace is today little remembered outside evolutionary circles.

Yet Wallace, this great ultra-adaptationist, this most Darwinian of Darwinians, did not believe that orthodox natural selection could explain our moral sentiments. Over his lifetime he, like both Darwin and

---

out, Darwin left enough room in *Descent* for accusations of both Lamarckism and group-selectionism. My point, though, is not that Darwin can confidently be said to have got the human animal right. It is that he still got a lot closer than the heretics who write today in his name.

A complete disregard

Huxley, was to become more and more convinced that something more was needed to explain the human moral sense. Even by 1864 he was noting that the early ancestors of man "had not yet acquired that wonderfully developed brain, the organ of the mind, which now, even in his lowest examples, raises him far above the highest brutes - at a period when he had the form but hardly the nature of man, when he neither possessed human speech, nor those sympathetic and moral feelings which in a greater or less degree everywhere now distinguish the race" With hindsight, in this passage Wallace remains more orthodox than Darwin. While Darwin saw very inchoate morality in some aspects of nature, Wallace, like the leading selfish gene theorists today, saw simply the amorality engendered by a process driven by reproductive logic alone.

Yet Wallace was to part company with Darwin and Huxley over the process that had raised man "far above the highest brutes". Darwin and Huxley both accepted that if you cannot invoke biology then materialism forces you to invoke culture. Wallace too pondered "the inverse problem" as he called it in his 1870 essay "The Limits of Natural Selection as Applied to Man": "an attempt ... to deduce the existence of a new power of a definite character, in order to account for facts which, according to the theory of natural selection, ought not to happen" (republished in Wallace [1891]). Wallace's conclusion (heavily influenced by a thriving nineteenth-century spiritualist movement) was that "a superior intelligence has guided the development of man in a definite direction", an answer that feet-on-the-ground materialists like Darwin and Huxley had no time for.

It is largely because Wallace turned away from earthly answers that he is today not better known. Wallace embarrassed Darwin with his public confessions about natural selection's inability to explain the most important human attributes, and through his seeming flight to spiritualism. "I hope you have not murdered too completely your own and my child," Darwin wrote anxiously to Wallace in April 1870. Wallace embarrassed Darwin, and he embarrassed the scientific establishment. However, it is precisely this turning away that may be Wallace's greatest achievement. In May 2000 the Harvard biologist Andrew Berry made the intriguing suggestion that Wallace perhaps may have been attracted to spiritualism partly *because* of his naturalism. As Berry noted, of all the nineteenth-century naturalists only Wallace had first hand knowledge of both the "lowest" humans and the "highest" apes, so possibly only he could have grasped how profound the inverse problem really was.

By turning to non-earthly answers Wallace did indeed get it badly wrong. But he got his answer wrong *because* he was capable of seeing the problem, *because* he espied the paradox at the heart of his own evolutionary theory. Many historians of science deride Wallace for his perceived unwillingness to apply natural selection consistently to mankind. But Wallace was actually one of the *only* scientists ever to apply natural selection consistently to mankind; one of the only scientists to openly address the idea that humans should have carried the same biological code as the rest of nature. So Wallace may have got his answer wrong, but at least he maintained his evolutionary integrity and kept his science intact. And he saw, all those years ago, that the test of a true Darwinian wasn't the answer you gave; it was in admitting to the problem in the first place: "But even if my particular view should not be the true one, the difficulties I have put forward remain".

Thomas Huxley
In his 1859 *Origin of Species* Darwin famously gave his cryptic comment on the implications of natural selection for human self-understanding. "Light will be thrown on the origin of man and his history", he wrote. Darwin had previously confided to Wallace that he would not deal with man's origin, even "though I fully admit that it is the highest & most interesting problem for the naturalist". Darwin was plagued by fears about the effect his theory would have on the stability of Victorian society, and by the reaction towards him from those whose respect he most hoped to earn. Many of Darwin's friends and colleagues were either within the Church of England, or very close to the Establishment, and by the time Darwin was forced to publish "transmutation" was still viewed with deep suspicion. Yet while Darwin would not at first make more than the briefest of comments about the most explosive application of all, this did not mean that he was not keen for others to take his ideas and apply them to "the highest & most interesting problem" for the naturalist. And the greatest of these proselytisers, the most uncompromising of his supporters, the zoologist who coined the term "Darwinism", was Thomas Henry Huxley. As Adrian Desmond and James Moore note in their excellent and much acclaimed biography of Darwin: "It was clear by April [*1860*] that Darwin's and Huxley's fates

A complete disregard

were now irrevocably entwined".

T. H. Huxley's contribution to both nineteenth-century Darwinian theory and British social transformation is impossible to overestimate. From evolutionary propagandist and creator of the concept of agnosticism, to educational and cultural reformer, Huxley's influence was felt everywhere. Without Huxley's Darwinian campaigning and agnostic polemics, the *Times* epitaph read in 1895, it would be impossible "to estimate the forces which have been at work to mould the intellectual, moral, and social life of the century". He shaped our vision, as Huxley's biographer, the historian of science Adrian Desmond, has written. And while it was to be more than a decade after the publication of *Origin* before Darwin turned his pen to man, Huxley came out fighting from the outset.

"Hurrah the Monkey Book has come!" Darwin wrote in February 1863 when Huxley's *Man's Place in Nature* was published. The book stripped away our idea of a noble past, offering man instead a lowly origin, but a noble future. But Darwin's Bulldog, as Huxley came to be known, was another who believed that the good in man cannot have arisen from evolutionary forces. Contrary to Darwin, who at least saw in nature a potential embryonic foundation for our subsequent culturally-enhanced morality, Huxley, like Wallace, saw not even the beginnings of a morality in the natural world. Huxley, the first Darwinian to research in detail our connection to the primate line, also saw clearly the brutality of our fellow apes.

For Huxley culture was necessary to oppose the legacy given to us by natural selection, the "cosmic process": "[l]et us understand, once for all, that the ethical progress of society depends, not on imitating the cosmic process, still less in running away from it, but in combating it. ... The history of civilization details the steps by which men have succeeded in building up an artificial world within the cosmos". "Social progress means a checking of the cosmic process at every step. ... Laws and moral precepts are directed to the end of curbing the cosmic process and reminding the individual of his duty to the community".

The current period - Williams, Maynard Smith and Dawkins

In one small sense, then, Cronin is not wrong to criticise Darwin's understanding, because Darwin was operating under a slight misapprehension. While Wallace and Huxley saw more clearly the moral indifference of nature, Darwin thought of nature as morally immature; something that needed building upon. It was to be another hundred years before nature was realised to be, not morally immature, not even morally indifferent, but downright malevolent. For morality to be possible nature had not so much to be built upon, as torn down so that one could start again.

I call people like Steve [Gould] and myself "the naturalists," in contrast with our gene-minded colleagues, the ultra-Darwinians. ... The three main characters in the ultra-Darwinian camp would be [John] *Maynard Smith, George Williams, and Richard Dawkins.*
Niles Eldredge – (in John Brockman's *The Third Culture*)

George Williams
The "selfish gene" revolution - the modern explanation of Darwinism that sees selection as predominantly operating at the level of the smallest unit, the gene - is generally held to have started in 1966 with the publication of George Williams' *Adaptation and Natural Selection*. Although Williams saw himself as continuing in a tradition first anticipated by major biologists such as R.A. Fisher, J.B.S. Haldane and Sewall Wright in work dating back to the early 1930s, it was not until Williams' clear analysis that the incompatible traditions began to diverge. In his book Williams developed "the formally disciplined use of the theory of genic selection for problems of adaptation". Unless evidence contradicted, the gene was to be recognised as the fundamental unit of selection, and Williams proposed that in evolutionary theory a gene could be regarded as any "hereditary information for which there is a favorable or unfavorable selection bias equal to several or many times its rate of endogenous change". This concept of the gene as any unit of developmental information visible to natural selection became the standard definition of the gene within the discipline, and the start of the "adaptationist programme" within Darwinism. (The terms "the selfish gene" and "selfish gene-ery" did not emerge until Richard Dawkins popularised the programme - and indeed made its position even more uncompromising - in his 1976 bestseller.)

Williams' book, subtitled *A Critique of Some Current Evolutionary Thought*, was also a criticism of theories of "group selection", or the idea that nature might select for the advantage of the group (even at the cost

A complete disregard

to the individual). Biologists had a long track record of appealing (sometimes only implicitly) to "greater good-ism" and the idea that nature may select for the good of the group, local population or species, but in 1962 a biologist named V.C. Wynne-Edwards had proposed that nature could and did select for the benefit of the group even at the expense of the individual. A serious attempt to resolve the group selection question was now unavoidable, and Williams began by reviewing the range of instances cited as examples of group selection. While noting that group selection was not impossible, he concluded that the adaptations in question could almost invariably be explained in terms of selection at levels lower than the group. Biologists did not need to appeal to group-level benefits where a lower level benefit explained the adaptation. Orthodox Darwinism has since been a continuously improving attempt to explain natural selection as operating at the lowest levels, that which is of benefit to the individual, or, more specifically, its genes.

Yet despite being the senior figure within selfish gene-ery, Williams has avoided the hostility that might perhaps be expected from the opposing "naturalist" camp of Gould and Eldredge. His careful style and intelligent, rather than blanket, criticisms have won the admiration of even those who profoundly disagree with the adaptationist programme. Indeed, Niles Eldredge's tribute to him in Brockman's *The Third Culture* is positively glowing. "George really is the most important thinker in evolutionary biology in the United States since the 1959 Darwin centennial. It's astonishing that he hasn't gotten more credit and acclaim. He's a shy guy, but a very nice guy, and a very deep and a very careful thinker. I admire him tremendously, even though we've been arguing back and forth for years now."

Yet "the most important thinker in evolutionary biology in the United States" is rarely asked his opinion of the human animal. And when he is, not for Williams the largely empty answers of the evolutionary psychologists[8]. George Williams' answer to the human animal is the same answer as Darwin gave, and it is the same as Huxley gave; in culture lies the key. As he first noted in a 1988 *Zygon* paper (and then

---

[8] Which is slightly ironic when you consider that Steven Pinker describes Williams' *Adaptation and Natural Selection* as "the founding document of evolutionary psychology" (Pinker [1997a], p. 56).

expanded a year later in a co-authored book), the answer to the human animal cannot lie in genetics, because nature is malign, not benign. There is no virtue in nature, only vice. Nature is our enemy, not our friend. For Williams, culture, in the form of powerful psychologically attractive worldviews and other motivators, "induces" a selfish ape to behave truly altruistically. This is cultural "manipulation" of our genetic inheritance: "[i]t is indeed correct that truly altruistic acts cannot be favored by selection, but the ability to induce others to behave altruistically certainly can be". He is happy to lend his name to "the Huxley-Williams nature-as-enemy idea", as he described it in his 1998 letter to me attached in the appendix. And what Williams realises, and as you will come to realise, is that if nature - if our evolved biology, our own genetic code - is "our" enemy, then something else is coming into play to make the difference. If nature is our enemy, then something else must be our friend; albeit a capricious friend.

John Maynard Smith
While George Williams is probably the leading name in professional selfish gene theory, the English evolutionary geneticist John Maynard Smith must run a very close second. In 1964 he was busy noting the difficulties with group-selectionist theories, including the damaging spread of the "'anti-social' mutations" Wynne-Edwards was denying would propagate. Today Maynard Smith continues to labour on the most fundamental questions within evolutionary biology. Yet his most valued work dates from the early 1970s when he became the first to apply game theory modelling to the natural world.

Game theory is a branch of mathematics developed by Morgenstern and von Neumann in the 1920s as a way of formalising social questions and activities, based on the underlying assumption of each player in the game acting rationally. An example of the way it works is as follows. The most famous game theory scenario is called Prisoner's Dilemma. Two prisoners, A and B, are awaiting trial and are faced with only two options. Each can "co-operate" with the other by refusing to confess to the crime, or else can "defect" by confessing to the authorities. There are only four possible outcomes. If both co-operate they will get only a light sentence (say 3 years) because the full evidence will not be available and so they can only be tried for a lesser offence. If both defect they will get fairly stiff sentences, but with some leniency for co-operating with the authorities

A complete disregard

(7 years). If A co-operates – refuses to confess – but B defects and gives evidence throwing the blame on A, then A will get 10 years and B only 2 years as a reward for helping the authorities. Correspondingly, if A defects while B co-operates then A will get 2 years and B 10 years. The point is that it always pays to defect in this game. Look at it from A's point of view. If A defects and *B has defected* then A is rewarded with a lighter sentence than the 10 years he would have got if he had stayed silent. If A defects and *B has co-operated* then A gets a much lighter sentence (2 years). It will not pay A to co-operate even when B co-operates because the 3-year sentence will not be as good. Since both are faced with these scenarios, both will always defect. So co-operation will not evolve in rational self-interested players. But, importantly, co-operation *can* evolve when the game is continuously repeated – known as Iterated Prisoner's Dilemma. Under these circumstances long term self-interest is best served by co-operative strategies of enlightened self-interest among players who continually have to interact. And this is exactly what happens in the natural world (where iterated natural selection has replaced the rational player), in the form of reciprocal altruism and delayed reciprocal altruism. Maynard Smith was to develop the idea far more deeply in his "evolutionarily stable strategies" (see later).

I first came across John Maynard Smith's human understanding in a 1988 collection of some of his earlier writing. This included a 1986 review of an attempt to explain human behaviour in terms of culture manipulating the genetic (by the ecologists Robert Boyd and Peter Richerson in their *Culture and the Evolutionary Process*) and where he admitted that: "I do not have anything better to offer". He had never been attracted by the work of the early human sociobiologists, because "human societies change far too rapidly for the differences between them to be accounted for by genetic differences between their members". In 1995 Maynard Smith was to openly support the "cultural manipulation" hypothesis as the key to human behaviour in a book he co-wrote with the chemist and physical scientist Eörs Szathmáry. The authors tell us that, when it comes to human societies, "[o]ften, the crucial factor is ritual ... and myth".

John Maynard Smith may well be as keen as George Williams to see an end to the harmful "Mother Nature" idea prevalent within human evolutionary biology for the last twenty-five years. When I had previously attempted to raise awareness of the vast yet almost unnoticed difference between the leading gene-selectionists and their more heterodox disciples in a paper submitted to *The British Journal for the Philosophy of Science* (see chapter 5), I had had to approach Professor Maynard Smith. I needed him to confirm for me (for my own peace of mind) that his more recent work was not, as had been suggested by the editors of the *BJPS*, a shift in his previous position. I explained that the point of my paper was not to contribute to the continuing debate between gene-selectionism and its critics, but was intended only to identify the very different traditions existing within gene selection itself. I told him that I was attempting to reconcile the one tradition represented by himself and Dawkins with my own background of political philosophy (at that stage I was only just getting into Williams' work). In an amusing letter back to me three days later he kindly confirmed my interpretation of his more recent work, and asked to see a copy of the paper "if it ever sees the light of day!" As I explain in chapter 5, that paper, perhaps unsurprisingly, never did see the light of day.

Richard Dawkins
Richard Dawkins, extremely contentiously, considers that culture may be a new form of "replicator" on our planet. In chapter 11 of *The Selfish Gene*, entitled "Memes: the New Replicators", Dawkins explained that most of what is unusual about man involves culture. Cultural evolution is not unique to man as, for example, certain birds can evolve new songs which they are not genetically coded for. But these are just "interesting oddities", he tells us, because in our own species the degree and speed of cultural evolution is staggering. And yet biological adaptationist suggestions for the degree and speed of that change "do not begin to square up to the formidable challenge of explaining culture, cultural evolution, and the immense differences between human cultures around the world".

So Dawkins suggested that man could only be explained by going back to first principles. All life for Dawkins evolves through the differential survival of replicating entities. Life on our planet evolved billions of years ago with the replicating entity of the gene. But can there be other kinds of replicators, and in consequence, other kinds of evolution, he asks? He suggests that a new kind of replicator has recently emerged on this planet. It is still in its infancy, he tells us, and is still drifting clumsily about in its primeval "soup" of human culture, "but already it is achieving evolutionary change at a rate that leaves the old gene panting far behind". This, for Dawkins, is the "meme", a unit of cultural transmission, a unit of *imitation*. Examples of memes are tunes, ideas, fashions and ways of building pots or bridges. The key for Dawkins is that memes propagate by "leaping from brain to brain" through the broad process of imitation. Memes survive and prosper in the meme pool depending largely upon whether they have great psychological appeal.

Dawkins' idea of the mental parasite has been criticised by many, but it cannot be easily written off as it has powerful supporters who also tend to see parallels with the study of biological parasites. George Williams himself has written on the utility of the idea, and in his 1998 letter to me he drew attention to manipulation by such mental "epidemics". Peter Medawar, the Nobel Prize-winning immunologist, and (with George Williams) one of the few evolutionary thinkers respected by all sides in the ongoing Darwinian debates, described Dawkins' memes departure as "his last and most important chapter" in his 1977 review of *The Selfish Gene*. The attraction of the idea to certain biologists is obvious; biological parasites infest their victims, they replicate inside their victims, they can have powerful effects on their hosts, but they can be combated. While accepting that mental parasites can never be the same as physical parasites, some notable biologists are therefore drawn to the understanding that the victims of both types of "parasite" need some form of initial susceptibility to infection; a run-down immune system, or a brain continually seeking solace.

That said, I personally think that the "meme" is one of the most harmful ideas to come out of evolutionary biology in decades. Not, I must quickly point out, because I think the idea is necessarily wrong. Indeed, up to a certain point I have much sympathy with the concept, and in chapter 9 we consider the very real problem of what "me" might mean separate from the dictates of those puppetmasters, genes and culture. It is this problem that has led a number of evolutionary thinkers to see culture in terms of parasites, Dawkins' "viruses of the mind". No; my hostility to the idea is largely because the concept tragically helped cloud the very issue it was intended to address. It is rather pointless to discuss the *form* of cultural manipulation until people understand what our genes tell us about the *existence* of cultural manipulation in the first place.

Social scientists have been critical of Dawkins' concept of the meme, hypothesised as analogous to the particulate gene. They have seen it as biologists trying to encroach on their territory, turning culture into little more than a series of unique viruses. Dawkins admitted in the second edition of *The Selfish Gene* that his "purpose was to cut the gene down to size, rather than to sculpt a grand theory of human culture". Yet the knee-jerk hostility of many established social scientists to, as well as the effusive support of a number of perhaps more radical social scientists for, the concept blinded almost everyone outside biology to what was going on here. Those who attacked the concept and those who supported the concept; none understood the crucial relationship it held to the gene. The crucial thing about the "meme" is *not whether the meme concept is valid, but the foundational genetic understanding it is built upon*. The meme, at least when it was originally posited, was viewed simply in terms of the "gap" it sought to bridge. Memes were simply "not genes" to begin with; they were what had to be posited when one began with the knowledge that we all start our lives as selfish apes. Dawkins, as an evolutionary geneticist, knew that genes could not come close to explaining human behaviour, and that something more was required. Maybe Dawkins is right to speak of the "meme", and maybe he is wrong; the point is that the meme fills a *gap*. A gap between the sorts of behaviour we should be expected to display as selfish apes, and the behaviour we actually do display.

As I have said, I have much sympathy for the idea that Dawkins is trying to get across, but to avoid unnecessary problems this book assumes we are only dealing with one true Darwinian replicator on Earth (the gene). I will avoid using this (currently) harmful concept of the "meme", and will simply concentrate on what genes tell us about ourselves, and what therefore (by elimination) they can tell us about what effect culture must be having. It is time now to turn to gene-selectionism, and time to understand why biologists, from Darwin, Wallace and Huxley to Williams, Maynard Smith and Dawkins, needed to bridge that "gap" in the first place.

The selfish gene
Evolution is a struggle for survival and works because it is. As George Williams clearly demonstrated in 1966, any adaptation that somehow managed to evolve for the good of the group would suffer from

A complete disregard

"subversion from within". Organisms within groups are still competing against one another, and the organisms that begin to take the benefits without paying the costs will be favoured by evolution. Those organisms that acted for the good of the group, rather than at all times being driven by what is good for themselves, would be taken advantage of and would not be the successful ones.

"Subversion from within" has consequences. John Maynard Smith made a major breakthrough in the early 1970s when he applied game theory to the natural world to develop the concept of the *evolutionarily stable strategy*, or ESS. These strategies are the possible group behaviours nature has discovered within the limits set by given genetic inheritance and reproduction systems. For example, natural selection has ensured that for diploid organisms close co-operation is only possible in small groups – for all other apes groups of around sixty to one hundred. An ESS is defined as a strategy which, if most members of a population adopt it, cannot be bettered by an alternative strategy. Such a strategy cannot then be bettered by an aberrant individual; the strategy is stable and cannot be subverted from within. No aberrant individual will have a higher fitness or potential for reproduction, and the trait thus becomes fixed. It is the paradigm of game-theoretic modelling, and where, of course, it is iterated Darwinian selection playing the role of the "rational" policy selector. "[N]atural selection provides a dynamics which will, subject to constraints, cause a population to evolve towards an optimum", as Maynard Smith noted in his concluding remarks in *Evolution of Social Behaviour Patterns in Primates and Man*. Groups are made up of competing individuals, and natural selection creates limits on group sizes, dependent on the level of co-operation necessary between the individuals, and the degree of relatedness amongst them. Biological theories of altruism work (in diploid/non-sterile caste species) for small family groups where there is a high degree of genetic relationship, and for small reciprocating groups where the parties regularly interact and can consequently recognise each other and remember the interchanges. Natural world "tit for tat" exchanges can only work under such a system. If one chimpanzee grooms another chimpanzee it is because it is looking for something in return from that other chimp.

Now group size is a function of many factors, including food availability and security. The type and frequency of prey may necessitate a larger group for regular successful kills, but this larger group means the disadvantage of having to share a kill amongst a larger number. Open areas require larger groups for protection of members against predators; in a wooded area smaller groups afford sufficient protection because individuals can more easily escape in the trees. Nevertheless, one crucial restraining *upper* limiting factor will be given by the degree of interaction necessary between group members. Wildebeest may sweep majestically across the plains in groups of many thousand, but wildebeest don't have to co-ordinate and share a kill and don't therefore have to work together. Behaviour that necessitates many interactions between group members requires strong group cohesion. This means individuals need to know each other well. Nature has established that for primates such cohesion cannot be left to degree of relatedness alone. Grooming is an essential part of primate behaviour because it allows members to form bonds and get into reciprocating relationships.

But group cohesion in primates is limited. Firstly, there is the need for strong reciprocating relationships that require detailed knowledge of your associates; detailed knowledge that can only be gained and maintained by forfeiting the possibility of a larger, more loosely knit structure. Secondly, grooming takes time. Primates may spend up to 20% of their time grooming for cohesion. They cannot afford more because over 80% of their available time must be spent hunting and working. So there are reasons for group sizes in nature. Group size is effectively kept "stable" by the positive and negative forces at play. And what do we see in our nearest neighbours? Chimpanzees live in groups of rarely above one hundred members. Even where we leave the apes and look at the less cohesive (and less closely co-operating) monkeys, we find that baboons will usually live only in groups of a couple of hundred. One species of baboon can live in groups of up to eight hundred, though what we always find with large groups in the primates is that they will be made up of numerous smaller sub-groups, with little peaceful interaction occurring between sub-groups. Yet humans coexist and interact peacefully in our *millions*. The only other animals capable of such a scale of interaction are the social insects, but as we now know, if we are looking for biological answers we cannot model ourselves on the social insects. There is no corresponding biological reason for why "we are in a sense the ants of the ape family".

Language on its own cannot account for the jump, because language

A complete disregard

is simply "grooming-plus". Language is an evolutionary tool that permits more efficient exchange of information and favours; it is still a product of natural selection and must therefore fit into the kin selection / reciprocity matrices of nature. Mutual grooming can be more efficiently replaced with a good gossip and still retain its group cohesion utility. Some biologists have estimated that to retain group cohesion by spending 20% of our time chattering we can account for groups of 150. Now this might be the size of your circle of friends, but in no way can it account for our larger group allegiances, to countries, parties, and religions. We are all so accustomed to going around our business and interacting peacefully with our "fellow men" (that staggeringly anti-Darwinian concept) that the ability to actually do so doesn't shock us. Yet the fact remains that ultimately we are products of natural selection, and orthodox natural selection cannot explain how groups of one hundred suddenly exploded into groups of thousands and millions.

"Gross immorality"
The defining feature of all evolutionary evangelism (to use Ridley and Dawkins' term) is the belief that nature selected man as "the decent ape", to paraphrase Wilson. Man, "the virtuous ape" (Matt Ridley). Man, "the moral ape" (Robert Wright). Destroying this idea that nature can select for morality is the point of this chapter and the next. You have seen the leading Darwinians rejecting the notion that, when it comes to human behaviour, it is genes that really count. You are probably a little curious to know *why* they reject the genetic answer so quickly.

One hundred years ago Thomas Huxley spoke of the "moral indifference" of nature. Over a decade ago George Williams rejected Thomas Huxley's term moral indifference and substituted his own term the "gross immorality" of nature, our "enemy". Why did he believe that Huxley had understated his case? "He knew little of the prevalence and violence of infanticide and cannibalism. ... Times have changed. Biologists today realize that unpleasant behavior can be important, and they are increasingly willing to study it." And what are the conclusions of the new adaptationist paradigm? "Simple cannibalism ... can be expected in all animals except strict vegetarians", Williams writes.

For selfish gene reasons that we shall develop later, all chimpanzees (both male and female) will tear apart the male infant of one of their females where the infant may have been fathered by a chimpanzee from another group. They will also kill (and sometimes eat) the male infant of a new female trying to join their group. The new female will attempt to protect her infant from the group up to a certain point, but after she has lost this one-sided battle she will not grieve for her dead infant. She will soon be willingly mating with her infant's killers. Forms of behaviour that go beyond mere moral indifference, and show us that the natural world is a world of gross immorality, that the natural world is the enemy of morality.

Gross immorality is the genetic programme of the other primates. Some, though, would prefer you to believe that humans evolved along an utterly unique genetic path. In the last chapter we saw the sociobiological explanations for human altruism, for our unique capacity for decency. For Wilson, large-scale human sacrifice exists because human evolution has worked in such a way that the species did better by evolving *unconditional, disinterested* altruism. Similar explanations come from the biologist Richard Alexander, who writes in *The Biology of Moral Systems* that "[p]opulation-wide indiscriminate beneficence might also evolve when small 'populations' are regularly composed of relatives related to a similar degree" and that this can thus be a basis for explaining human large-scale generosity. This argument is also proposed by David Barash in *Sociobiology and Behavior* when he says that "unilineal descent permits extraordinarily large numbers of individuals, from dozens to millions, to unite under the aegis of common ancestry". Many of the early big-name human sociobiologists also based their explanations for why we are programmed to be "nicer" than other animals on the suggestion that we evolved the capacity for third party reciprocal altruism, where returns may thus eventually come back from society at large. And evolutionary psychology theories of human altruism are little changed from the earlier tradition they base themselves on; positing newly evolved "conscience modules" in our brains, or simply fortuitous genetic expression.

Darwin realised that human morality is not a natural kind. Make no mistake; his conclusion grieved him. As a proud and privileged Victorian largely intent on maintaining the stratified social ordering of his time, he would have been very keen to place human behaviour predominantly within biology. But Darwin was first and foremost a scientist, and he could not reconcile the natural world with the human world. Darwin knew that to argue that modern human behaviour was largely biological

A complete disregard

would have been to betray his own theory of evolution by natural selection. Darwin knew that to suggest that humans do not share the same basic biological programme as the remainder of the natural world would be an argument based on anti-Darwinian heresy.

3

Natura non facit saltum

On the theory of natural selection we can clearly understand the full meaning of that old canon in natural history, "Natura non facit saltum".

Charles Darwin – On the Origin of Species

"Natura non facit saltum". "Nature does not make a leap." It is time to consider in depth the heresies of those who suggest it can.

Problem 1: "cultural evolution, and the immense differences"
The first indication that there may be something seriously wrong with the "all-in-our-genes" worldview was simply a logistical problem. Darwinism works over huge timeframes. It works over many thousands of years, even over millions of years. Darwinian inheritance is about the ever-so-gradual accumulation of small changes (a point we shall return to below). Yet human cultures change so incredibly swiftly, with the divergent behaviours noted by philosophers from Aristotle to Nietzsche taking place not just across isolated groups, but within those same groups over relatively short periods of time. It was the amazing speed

37

with which human groups could move from displaying one typical characteristic to, just a few generations later, displaying its polar opposite that made many philosophers realise that human behaviour probably had little to do with biology. And even more so than philosophers, the leading evolutionary geneticists realised this was a major problem for any genetic explanation. "[H]uman societies change far too rapidly for the differences between them to be accounted for by genetic differences between their members", writes Maynard Smith in *Games, Sex and Evolution*, a collection of some of his earlier writing. The paper from which the quote was taken was written in review of Wilson's *On Human Nature*, and the timeframe problem was one of the main reasons Maynard Smith rejected Wilson's suggestion that biologists can add anything material to human self-knowledge. A few years later Wilson was to publish his book *Genes, Mind and Culture* together with the mathematician Charles Lumsden. This book, which "allowed" a greater evaluation of the role of culture than Wilson had previously considered, was based around the thesis that, as culture changes rapidly, populations (through differential survival effects) could acquire a genetic predisposition towards those new ways of life. The point was to show that his earlier work had not been wrong to largely ignore possible effects from cultural pressures. Genes hold culture on a leash, as superior genotypes are favoured across the progressive cultural timescale. Again, Maynard Smith, as a trained population geneticist, was forced to point out the flaws in the genetic models. "[T]he mathematical models ... fail to demonstrate any synergistic effect between cultural and genetic processes. ... On three occasions Wilson has found it helpful to find a mathematical collaborator. ... [H]e was third time unlucky." (Maynard Smith's review of *Genes, Mind and Culture* is also reprinted in *Games, Sex and Evolution*.)

Richard Dawkins makes this point about the slow speed of genetic change relative to the varied histories of human cultures equally forcefully in *The Selfish Gene*. He notes that there are some similarities between natural world behaviour and human behaviour (although, as Dennett warned us earlier, this by no means implies there is a genetic basis to the similarity). Monogamy - and more specifically a female's determination to withhold copulation until the male shows some evidence of long-term fidelity - is a strategy found often in nature. "Many human societies are indeed monogamous. In our own society, parental investment by both parents is large and not obviously unbalanced. ... On the other hand, some human societies are promiscuous, and many are harem-based. What this astonishing variety suggests is that man's way of life is largely determined by culture rather than by genes". The explanations of the human sociobiologists "do not begin to square up to the formidable challenge of explaining culture, cultural evolution, and the immense differences between human cultures around the world", he continues.

As Daniel Dennett reminds us in his *Darwin's Dangerous Idea*: "virtually all the differences discernible between the people of, say, Plato's day and the people living today - their physical talents, proclivities, attitudes, prospects - must be due to cultural changes, since fewer than two hundred generations separate us from Plato".

Problem 2: A universal nature

An even more fundamental problem with much human sociobiology (and *all* behavioural genetics) is universality, or rather the lack of it. Along with our common genetic predispositions, Edward Wilson had also wanted to show that individual differences in behaviour are largely the result of the slightly different genetic codes we are born with. With the exception of identical twins, no two human beings carry the same genetic code, the same genome. Couldn't we therefore argue that the great differences in human behaviour are the result of the slight differences in our genetic make-up? Unfortunately for Wilson, *no*, we could not, although it took almost a decade for this very simple evolutionary point to sink in. And sink in it eventually did. Evolutionary psychology severs almost all contact with the earlier human sociobiology because it so desperately does not wish to be seen making this same mistake. What mistake?

Most human sociobiology wanted to prove that humans behave differently because we all have slightly different genotypes. It totally ignored the point that all species produce individuals with slight differences in their genetic codes (this, indeed, is why evolution has something to work on). While the level of genetic variation depends upon the species, statistically any individual in any animal species will differ from any other member of its own species chosen at random by at least 0.1% of its DNA. There is more genetic variation in orang-utans, or lions, or wildebeest, than there is in humans, simply because humans evolved so recently that there has been little time for variety to creep in. *And yet all orang-utans and lions and wildebeest behave in effectively the same very predictable ways.* The point that Wilson had entirely missed was that evolutionary inheritance has produced *universal* animal natures. So what does the genetic variation account for? For small differences in body design or function, not large differences in behaviour, that's what. Natural selection creates species-wide behaviour patterns. It might create some animals within a species that can run a little faster, or see a little better; but fast or slow, keen or not so keen of vision, they have the same rather predictable ways of behaving. Ways that you have begun to understand, and that you shall meet again in all their glory in the next chapter.

Nature is about gross immorality. But it is more than that; it is about *species-wide* gross immorality. *All* chimpanzees will tear apart the male infant of one of their females where the infant may have been fathered by a chimpanzee from another group. There is not one, or a small group, sitting on the sidelines wailing "No! Stop that!" There are not other chimpanzees sitting in a bordering territory watching *The Jungle's Most Wanted* and decrying the state of infanticide in their neighbours. Natural selection gave all chimpanzees the same general way of behaving. This is not to say that all chimpanzees behave identically, because infanticide requires environmental triggers, such as an unwelcome infant, and will also depend on other contingencies such as group position. But when those triggers are around, the entire group (males and females) will co-operate in "removing" that unwanted infant. Females can form coalitions against infanticidal males, but only when there is strong relationship between the females and the females tend to face an identical threat posed by a newly dominant male. Nature is about species-wide ways of behaving, because the selfish gene has the same ruthless logic for all its products. A female that will today be running from infanticidal males will tomorrow be helping her group pull apart another's infant.

Natural selection bequeathed human beings the same species-wide way of behaving, too. You can try to tell stories about how humans received a *different* behavioural code from the rest of nature if you like (and as evolutionary psychology does). But if you want to posit genetic control of behaviour, *this* is your starting point: our species *shares* a biological behavioural code. It was a mistake evolutionary psychology vowed never to be caught making. This is why, as evolutionary psychologists are desperate to tell you, their "default assumption" is (purportedly) that we share a universal human nature; that our similarities may be genetic, but our differences are cultural.

Problem 3: One very lucky ape
No one will ever be sure how many species have existed on this planet since life began almost four billion years ago, but it is now thought to be many billions (remembering that 99% of all species are extinct). Let us therefore be conservative and say two billion. The rest of the natural world is genetically coded for what Williams termed "gross immorality" (what this really means we shall see next chapter; for the time being please just accept it probably doesn't look good). What both human sociobiologists and evolutionary psychologists need you to believe is that nature selected as follows:

Total species:	2,000,000,000
Grossly immoral species:	1,999,999,999
Virtuous species:	1

Now the point about statistics in biology is that probability factors depend on how you classify your data. Every species has evolved only once, and therefore every species is, in a sense, a statistical improbability. Well *yes*, if you are talking about the probability of evolution having created just this animal with its specific combination of particular features, one tiny offshoot of the great tree of relatedness. But emphatically *no* if you are talking about a type of animal. The contingencies that produced an African lion looking exactly as it does are unique. However, the probability that nature would create some animal to occupy a lion's particular evolutionary niche, and that the resulting animal would be somewhat similar to a modern lion, is relatively high. This is because the lion is rather well-adapted to that niche, and anything occupying such a niche would tend to have somewhat similar qualities (four legs for speed, strength, powerful jaws, similar level of body fat and fur to retain and radiate heat, similar colouring for heat loss plus camouflage, and so on).

But in hypothesising man, this is not what the evolutionary psychologists do. They do not work to the existing template, citing tiny

Natura non facit saltum

cumulative adaptations, each of which is a tiny change to an existing design, a small branch on the tree of evolution. What they do is throw away the template, and argue for a *qualitative* change, not simply a *quantitative* improvement. They are not saying we are chimpanzees plus large brain, minus body hair, plus upright posture. They are not saying we are "gross immorality-plus". When it comes to man they are rejecting the entire template, and throwing away everything that went before. They are tearing up the rulebook, and inventing a *type* of creature *utterly* different from its 1,999,999,999 brothers. Not just another twig on the great tree of life, but a whole new type of tree.

Evolutionary psychologists desperately want you to believe that you are (genetically) not like those other *nasty* apes. So *statistically* this hypothesis is therefore vastly improbable. But *mechanically* there is simply no process by which it could have occurred, as we shall see below.

Problem 4: When size does matter
We have already touched briefly on this problem. Common chimps can live only in groups of up to one hundred or so individuals, and pygmy chimps tend to live in slightly smaller groups. Yet humans live side-by-side with non-kin in cities and nations of *millions*, group sizes unknown outside the immense kin groups of the social insects. The philosopher of science Elliott Sober and the biologist David Sloan Wilson are no friends of selfish gene theory. Yet, like the leading selfish gene theorists, they, too, openly admit their surprise that humans can live in cities. Human groups have, like social insect colonies, "been interpreted as superorganisms for centuries", say Sober and Wilson in their book *Unto Others*, but biologists need some explanation for "why humans are ultrasocial". Sober and D.S. Wilson are arguably the two most influential levels-of-selection theorists within what is known as the "multilevel selection" tradition which opposes the gene selection tradition of Williams and Dawkins. Sober and D.S. Wilson, therefore, find their answers to human cohesion largely in group selection. As they put it: "At the behavioral level, it is likely that much of what people have evolved to do is *for the benefit of the group*" (the emphasis is theirs). Selfish gene theorists have no such easy outlet, however.

Altruism in nature can, we have seen, be explained through the two main mechanisms known as reciprocal altruism and kin selection. It was

Bill Hamilton who first drew the biological community's attention to the importance of kin selection in his seminal two-part paper in 1964. "Hamilton's Rule" as developed in that paper was applicable to all species by virtue of their relatedness, but the paper is particularly remembered for the way it was applied to the genetical asymmetry of haplodiploid Hymenoptera (which includes ants, bees and wasps, but not termites). Mammals, being diploid, have a double set of chromosomes, one from each parent. In haplodiploid Hymenoptera females develop from fertilised eggs and have a double set of chromosomes, while males develop from unfertilised eggs and have only a single set of chromosomes to pass on (they are haploid). Sperm from a male are thus genetically identical. The coefficient of relatedness of mother to daughter has the normal value of 0.5. But the average relatedness between daughters from this male is 0.75, closer than it would be to any offspring they might conceivably have. Helping your siblings - who may number in the hundreds of thousands - can become of overriding genetic importance. Hamilton noted that family relationships in Hymenoptera are potentially very favourable to the evolution of reproductive altruism.

Another route very favourable to the evolution of reproductive altruism is inbreeding because relatedness can rise above the value of 0.5 that applies under outbreeding. Thus, noted Hamilton, an individual should be more altruistic than usual to its immediate kin. Termites, for example, are not haplodiploid, but indulge in cycles of intense inbreeding within colonies and outbreeding to found new colonies. There are subtle problems to seeking the origin of social insect eusociality in simplistic application of coefficients of relatedness, and the question of insect eusociality is therefore by no means settled. Nevertheless, modern biology seeks its answers to the origin and maintenance of extreme insect sociality within (and blending) a limited range of possibilities, from very close genetic relatedness, to pheremonal suppression of fertility, parasitic influence, resource dependency and other ecological factors. Vast offspring production plus irreversible reproductive specialisation, morphologically delineated castes and a high level of hard-wired behaviour may also enable social insects to achieve group sizes impossible within the non-human mammals.

Important to this last understanding is the naked mole-rat. The naked mole-rat is "arguably the closest that a mammal comes to behaving like social insects such as bees and termites, with large colonies and a behavioural and reproductive division of labour", write Bennett and Faulkes in *African Mole-Rats: Ecology and Eusociality*. The Damaraland mole-rat is the only other eusocial mole-rat, yet colonies are tiny, with an average size of only 11 members, and a maximum size so far found of 41 members. Damaraland mole-rats have a strong inbreeding avoidance mechanism and have a relatedness coefficient of no higher than the normal 0.5 found in outbred first degree relatives. In contrast, even more extreme eusociality is displayed by the naked mole-rat, where colony average size is 80, and colonies of almost 300 animals have been discovered. Intense inbreeding among naked mole-rats has led to average intra-colony relatedness of 0.81, the highest recorded for a natural mammalian population. Comparison between the two species and other mole-rat species has tempered the initial view that it is inbreeding that explains all in naked mole-rat eusociality. Though levels of relatedness are always an important factor, ecological factors are also now thought to be critically important. Harsh and unpredictable climate and location producing benefits from assisted living, and very high costs constraining dispersal (both in terms of creating a new subterranean environment from scratch and low probability of finding alternative food and mates) are implicated.

There are reasons for why social insects display their extraordinarily high degree of co-operation. There are also reasons for why naked mole-rats display their kin directed co-operative breeding. Yet these models are so very different from what occurs in *Homo sapiens*. There is no suggestion that unique genetic, or fertility, or parasitic, or ecological factors have played a role in human social development. Every leading selfish gene theorists realises that biology cannot explain human cohesion. As Maynard Smith says of altruism in *Games, Sex and Evolution*: Biologists "have explanations - such as the fact that the altruist may share genes with the recipient of its altruism, and it is genes, not individuals, that matter in evolution - but they are ones that work only for altruistic behaviour among the members of small groups".

Human cohesion cannot be explained biologically. Genetically, we should not be able to live in groups any larger than those of the other apes; groups of one hundred or so individuals. We should not be able to live in cities. Evolutionary biology can explain why animals, such as apes, that share fifty percent of their genetic material with their brothers and sisters can live in groups of one hundred or so. Evolutionary theory can also explain why creatures such as ants, which use unique genetic inheritance and reproductive systems to raise that level of shared relatedness to well above fifty percent, can live in groups of millions. But what evolution cannot explain is why humans, who share their inheritance and reproductive systems with apes, can live in groups of millions.

Human groups have, like social insect colonies, "been interpreted as superorganisms for centuries", say Sober and D. S. Wilson. This is Huxley's Paradox. This is the paradox of man. Genetically, we are apes. Yet we live as ants.

Problem 5: "implicitly group selectionist"
*As an enthusiastic Darwinian, I have been dissatisfied with explanations that my fellow-enthusiasts have offered for human behaviour. They have tried to look for "biological advantages" in various attributes of human civilization.
… Frequently the evolutionary preconception in terms of which such theories are framed is implicitly group-selectionist.*

Richard Dawkins – The Selfish Gene

As we have seen, Charles Darwin was an individual-selectionist. "[N]atural selection can act only through and for the good of each being. … Natural selection … will adapt the structure of each individual for the benefit of the community; [but only] if each in consequence profits by the selected change." Today selfish gene Darwinians reinterpret natural selection as operating almost exclusively at the level of the gene, or "gene selection".

What Darwin and modern selfish gene biologists both tend to strongly disavow is "group selection", or the idea that nature will select for the benefit of the group, rather than its component parts. Any benefit to a group can only be as a by-product of what is good for its individuals, and even then the behaviours are ruthlessly controlled. When Helena Cronin turned her Ph.D. thesis cataloguing the breakthroughs of twenty-five years of gene selection biology into her book *The Ant and the Peacock*, group selection was singled out for criticism. As Paul Griffiths wrote in *The British Journal for the Philosophy of Science* in his 1995 review article "The Cronin Controversy": "Group selection is the villain of much of Cronin's book". George Williams pointed out to me that I was probably asking for trouble when I made the comment in *Philosophy* that group selection "will not occur" in nature. I should settle for noting that group selection will usually be too weak to produce noteworthy effects (because of the much stronger and often counteracting forces operating at the lower levels). To say that group selection will not occur would invite the unnecessary additional hostility of the tradition associated with the late Stephen Jay Gould. Williams has been one of the few wisely trying to avoid unnecessary hostility over this point (as well as giving credit where credit is due to Gould and those in his alternative tradition).

The reason that gene-selectionists hold that group selection will not usually have notable effects is that evolution is a struggle for survival and works because it is. Any selected useful trait that managed to evolve for the good of the group would suffer from "subversion from within" because organisms within groups are still competing against one another, and the organisms that begin to take the benefits without paying the costs will be favoured by evolution. Those organisms that acted for the good of the group, rather than at all times being driven by what is good for themselves, would be taken advantage of and would not be the successful ones in the struggle for existence (see chapter 5 for a slightly more detailed explanation). And yet, time and again, group selection is what sociobiological theorising boils down to, as Dawkins points out above. Those who posit "biological advantages" to human behaviour can only do so by abandoning the rules of selfish gene inheritance. Group selection is anathema to mainstream Darwinism, yet here is Dawkins telling us that the sociobiologists and evolutionary psychologists - who profess to be good gene-selectionists - are actually heretics. This idea that we evolved to be "the virtuous ape" is nonsensical to the biologists who know that such a process could not get past the "subversion from within" problem at the very least. This is ironic because it has often been these heretics who have been most vocal in their criticism of other group selection theorists (such as Gould). This is especially true of the pop sociobiologists; the science writers and journalists. Pick up any of their works and you will invariably find a long section on "The Group Selection Fallacy", or some similarly headed piece. And yet it tends to be the popular sociobiologists who are most openly group-selectionist, and particularly our popular

Natura non facit saltum

proselytisers.

Let us return to John and Mary Gribbin. Birds, we are told, have evolved true altruism too: "[t]he pattern of behaviour that makes for helping at the nest of your siblings is almost exactly the same as the pattern of behaviour that makes for helping at the nest of any member of your species. ... [H]elping becomes a common activity ... as genes for helping spread through the gene pool" they write in *Being Human*. Unfortunately "general 'niceness'" is not the distinguishing feature of birds, for very important evolutionary reasons. Murderous indifference is. It actually tends to be birds that are used as models of efficient nastiness by the leading gene-selectionists. Dawkins tells us about blackheaded gulls within five pages of beginning *The Selfish Gene* and as a paradigm of gene-selfish behaviour. "It is quite common for a gull to wait until a neighbour's back is turned, perhaps while it is away fishing, and then pounce on one of the neighbour's chicks and swallow it whole. It thereby obtains a good nutritious meal, without having to go to the trouble of catching a fish, and without having to leave its own nest unprotected". Birds are no more altruistic than any other vertebrate. They cannot be; even were – and contrary to the rest of the natural world - such an originating mechanism somehow capable of inducing widespread altruism, subversion from within would automatically begin to act against it. Equally crucially, birds are no more vicious than other animals (as we shall see). The reason why such bird behaviour is "quite common", to use Dawkins words, is partly that they have more opportunity for such behaviour given the problems with fixed nesting sites, partly that much detailed observation has already been done on birds because 24-hour access is relatively easy, but mainly *because they are products of nature*. In a previous chapter we discovered that true "niceness" in birds is perhaps even a mutated form of kin selection for the Gribbins. But such an hypothesis runs directly counter to the last thirty years of orthodox evolutionary biology. Should such a mutation occur, it would immediately be selected against. Birds without this coding for selflessness would leave more offspring than those with such a coding. "Niceness" would be bred out of the population within a very few generations.

"[S]uch a tendency then begins to act 'for the good of the species', although it has its origins firmly in the selfishness of genes". No, under the logic of natural selection *it does not*. Nature has been viewed through rose-tinted spectacles for many hundreds of years. As Richard Dawkins tells us in *Unweaving the Rainbow*: "The medieval bestiaries continued an earlier tradition of hijacking nature as a source of moral tales". He lambasts the "spate of authors" today reacting indignantly to the idea that nature is genetically selfish, singling out the primatologist Frans de Waal, author of the tellingly-titled *Good Natured*. De Waal's open hostility to selfish gene-ery is apparent when he says of an understanding that explains evolved altruism entirely through selection at the level of the gene: "A more cynical outlook is hard to come by". He entirely misses the paradox first spotted by Wallace and then developed by Huxley (a paradox that existed even when selection was at the level of the selfish individual). John Gribbin's claim that "helping the sick and the weak ... emerged through the process of natural selection" seems extraordinary when even de Waal must admit that "there is no shred of evidence that other monkeys have ever gone out of their way to assist [*Mozu, a well-filmed monkey who has survived in a Japanese National Park despite severe deformities*] in her monumental struggle for existence". Yet de Waal refuses to see the paradox, describing as "monumental confusion" the position of those like Bill Hamilton, "the discoverer of kin selection, [*who*] has written that 'the animal in our nature cannot be regarded as a fit custodian for the values of civilized man'". Unfortunately de Waal, like so many others, seems hostile to selfish gene-ery largely because of his incorrect assumptions about what it must imply for human behaviour ("this view is ... enough to give goose bumps to anyone with faith in the depth of our moral sense"). It is a shame, though, that Dawkins has been so quick to attack group-selectionism in those who come from alternative Darwinian traditions, while doing so little to end the implicit group-selectionism often found in his supporters with their own models of animal "genuine kindness". While de Waal is torn limb from limb for daring to suggest that even non-human animals may have evolved unconditional altruism and good natures, Dawkins goes remarkably quiet when putatively gene-selectionist human sociobiologists and evolutionary psychologists suggest that the human ape, or even birds, evolved unconditional altruism and a good nature.

Time and time again, genetic group selection is what evolutionary psychology ultimately boils down to. In 1992 and 1993 there was an acrimonious exchange of letters in *The New York Review of Books* between Stephen Jay Gould on the one side and John Maynard Smith and Daniel Dennett on the other. Gould had been asked to review Helena Cronin's *The Ant and the Peacock*. While his review was unnecessarily hostile, he was trying to make an important point about her speculations on the human animal. Gould wrote in his review: "[o]r we might argue, as Cronin often suggests, that our general altruistic urges evolved long ago by kin selection. ... True altruism to nonrelatives would then be a consequence of formerly advantageous behaviour, now altered by a changing social setting that makes neighbors of genetic strangers ... [A]n untestable speculation". Elsewhere Cronin has tried to argue, with Matt Ridley, that reciprocity is the root of human kindness. "Now reciprocal altruism can be the font of *vast* altruism, self-sacrifice and *genuinely* societal values", she has argued (and I use her spoken emphasis from BBC Radio 4's "Basic Instincts", *Analysis*, 15 May 1997). Yet Cronin's hypotheses are utterly invalid. You cannot simply take mechanisms found in nature and suggest that they can be used to explain behaviour on a totally unique scale, or of a totally unique degree, because nature subverts down to small-scale minimum equilibrium. As Maynard Smith reminded us earlier, biological explanations of altruism work - in diploid, non-inbred species - only for the altruistic behaviour among the members of small groups. To claim that reciprocal altruism or diploid kin selection can be "the font of vast altruism" is to ignore the last thirty years of selfish gene Darwinism. To claim as Cronin, Ridley and Gribbin do is to be "implicitly group-selectionist".

Problem 6: The maladaptive ape
In *The Language Instinct* Steven Pinker takes great pains to note that evolution is all about creating universal natures. Just as with our universal capacity for language acquisition, "there is a universal design to the rest of the human mind".

We saw in an earlier chapter that evolutionary psychology often trades in unfalsifiable Just So Stories. This is not its real problem, though. Its real problem is that it trades in such poor Just So Stories. Remember the "adaptation-executors" argument? Virtuous behaviour can be accounted for because we are "adaptation-executors" and not "fitness-maximisers"? As such unconditional altruism never evolved as such, but it is the evolutionary consequence of 100,000-year-old selfish behaviour that in today's alien environment has the manifested expression of virtue and selflessness? The problem is you have to find a way to tell a story that gets you from the "gross immorality" of the natural world to the "morality is a biological adaptation" of the evolutionary psychology world. How do you do this? Human sociobiology simply threw the Darwinian textbook out the window and argued that we evolved a genetic code unrelated to that found in our closest primate kin. Evolutionary psychology, however, learned some very important lessons from the mistakes of human sociobiology; namely that nature codes for species-wide behaviour (see Problem 2), and that you just can't throw away the gene-selectionist rulebook (Problems 4 and 5).

So when it came to finding their own stories for how a product of a grossly immoral process ends up displaying morality and virtue, evolutionary psychologists were (initially) very wary of positing genetic *evolution*. Instead they posited phenotypic evolution: utterly unique expressions of that still grossly immoral biological make-up in novel environments[9]. So when our ancestors moved from the savannahs of Africa to the skyscrapers of New York, their "hunting wild game" genes turned into "playing the stock market" genes. Now this looks only slightly strange when imagining the change from hunting on the plains to surfing on the financial ether. But it looks remarkably funny when trying to change from a cannibalistic ape to a vegan hippy by positing the novel expression of those same cannibalistic genes.

A wonderful example of how strange this can look was given by the evolutionary psychologist Lewis Petrinovich recently in his *The Cannibal Within*. Petrinovich argued that cannibalism is an adaptation in *Homo sapiens*, just as cannibalism is an adaptation in chimpanzees. Humans do therefore carry the genetic code for cannibalism, it is admitted, *but*, Petrinovich argued, this is not our *true* evolved nature, this is not our *real* genetic code. Humans, uniquely, evolved to be moral, and cannibalism therefore only rears its ugly head when our older, baser, instincts resurface in times of starvation and other such stresses. In other words, when we revert to the primal code, to our ancestral ape nature. So, just as a drowning man may try to force another under in his terror as he begins to lose his humanity, a starving man may also revert to his primitive, pre-human form. Because, as the

---

[9] The phenotype is the outward expression of a gene. It is how a gene manifests itself in the actual body design created by the genes in interaction with the environment.

anthropologist William Arens once wryly noted, "real" human beings do not eat each other: "[E]ating human flesh implies an animal nature which would be accompanied by the absence of other traits of 'real' human beings".

Evolutionary psychologists, desperate to claim hegemony for the gene, must therefore either cling, like Petrinovich, to the belief that our grossly immoral genetic code does not really express itself (while other, never before identified, group-selectionist patterns such as "conscience modules" do), or to the belief that it does express itself, but has a unique genetic expression. The idea that our ancestors' most immoral genetic traits may not express themselves shows an optimism (and lack of evolutionary knowledge) that is touching. Gross immorality, as we shall see below, is not the code of just one or two genes; it is the code of a species' entire genome. Yet the alternative - that these genes do express, but have a fortuitously unique expression - is even stranger. The suggestion now is that the genetic code for cannibalism – which, remember, "can be expected in all animals except strict vegetarians" (George Williams) - just happens to have the phenotypic expression in modern man of, perhaps, the love of a good curry, or, in the case of our hippy friend, a lentil casserole.

Problem 7: Hopeful monsters

Darwin was a passionate anti-saltationist, and this led him to stress, over and over again, the extreme gradualness of the evolutionary changes that he was proposing.

Richard Dawkins – The Blind Watchmaker

"Natura non facit saltum" - "Nature does not make a leap." This is an understanding that pre-dates Darwin. The Latin expression itself quoted by Darwin in the sixth chapter of *Origin* was from Linnaeus' classic 1751 work on taxonomy. "Saltation" theories are theories that rely on macromutation, or a sudden large beneficial jump that is consequently incorporated into the gene pool of a species. "Saltationism" is a neo-Darwinian heresy.

And yet evolutionary psychology's claims are not simply "implicitly group-selectionist" involving processes impossible under Darwinian inheritance, they are also saltationist. The heretics' theories would fundamentally have to re-write natural selection's genetic code; the

Natura non facit saltum

gross immorality spoken of above and that we shall discuss in detail later. Evolutionary psychologists accept that up until 100,000 or so years ago our ancestors were selfish apes. So, if the evolutionary psychologists were to be right, what would have to be the implications for that last genetic change?

All primates are born with an effectively identical behavioural genetic code, the code George Williams calls gross immorality. This codes for species-wide patterns of cannibalism, infanticide (and the voluntary mating with those animals that have killed your infant), rape, levels of lethal violence against same-species members many thousands of times higher than rates found in even the worst human societies, and absolute indifference to the suffering of non-kin. Patterns not that *some* within a species are coded for, but that *all* within a species are coded for. Such conclusions followed from numerous studies. This code is not restricted to a few murderous genes; gross immorality is the programme of natural selection. This is a programme that has been favoured by selection over billions of years because it works extremely efficiently from an evolutionary point of view. Utter selfishness and indifference to others' suffering (where they can be of no use to you) is a message coded into the entire natural world behavioural genotype, it is not something tacked on a few million years ago. It has been the *guiding force* behind evolution for almost four billion years, and it dominates all behavioural coding. Gross immorality is not simply a characteristic of natural selection: gross immorality *is* natural selection.

Nature never starts again, and can only build on what has gone before. In *Plan & Purpose in Nature* George Williams introduces us to what biologists call *phylogenetic constraint*. "Evolution never designs anything from scratch. It can only tinker with whatever happens to be already there, saving slight modifications that provide immediate benefits, culling those that cause harm. Much of anatomical human nature derives not from anything currently desirable but from adaptive changes made in the early history of the vertebrates." We have two pairs of limbs, says Williams, not for functional reasons but for purely historical ones. "[T]he first lungfish that crawled from the water and pushed its way through the mud did so with the help of two pairs of appendages." Degeneration, too, is an historical drama, albeit one played out at a different speed. Snakes are limbless - though some do have claw-like hind limb remnants - as their limbs were gradually selected to get smaller over time (snakes have also lost eardrums and moveable eyelids, but gained vertebrae). In general degeneration can be enormously faster than generation. It required millions of years to make vertebrate eyes, but only thousands of years to lose them in caves. And as historical constraints exist with regard to anatomy, so they exist with regard to behaviour, for similar reasons. Each part of the anatomy, and each basic natural world behaviour, is coded for by numerous genes and linked to other parts of the anatomy and other behaviours. By changing one factor it may have vast implications for many other factors; and nature has no foresight. Nature progresses by the efficient selection of purely random mutation. Nature has to work within the limitations given by the fact that existing features are already coded, and de-coding is therefore greatly constrained. You cannot simply ignore what has gone before, or pretend it does not matter. Nature has taken millions upon millions of years to code for the set of behaviours in primates. While you can tinker with these behaviours, and add to them, and remove certain behaviours gradually over time, you cannot rip up the rulebook and start again.

The zoologist Jared Diamond wonderfully described humans as "the third chimpanzee" in his book *The Rise and Fall of the Third Chimpanzee* published in 1991. Humans share around 98.5% of their DNA with both the common chimpanzee (*Pan troglodytes*) and the pygmy chimpanzee (*Pan paniscus*, and also called the bonobo). We split off from a common ancestor with the chimpanzee[10] and the bonobo some six or so million years ago. The point of the first part of Diamond's book was to show (using the work of the molecular biologists Charles Sibley and Jon Ahlquist whose studies have since been reconfirmed by others) that human beings are actually more closely related (genetically) to both species of *Pan* than both species of *Pan* are to the other apes[11]. As Diamond noted, the common ancestor to the chimp, the bonobo and the human split off from the ancestor to today's gorilla (the next nearest

---

[10] Using the convention of primatologists like Frans de Waal, I will use the term chimpanzee to mean *Pan troglodytes* and the term bonobo to refer to *Pan paniscus*.

[11] The human / chimp relationship is just under 99% when calculating using single nucleotide substitutions. This method has been standard when comparing human / chimp DNA, but can be slightly misleading. For example, in September 2002 Roy Britten reported that the human / chimp relationship falls to perhaps 95% when one also

ape) more than a million years before the common ancestor to chimps, bonobos and humans split. In consequence, we and the chimp and the bonobo all share just less than 98% of our DNA with the gorilla. And the genetic distance separating humans from bonobos or chimps (around 1.5%) is actually less than the genetic distance between the common gibbon and the siamang gibbon (who were found by Sibley and Ahlquist to have a 2.15% variance). Our DNA is remarkably close to all other apes, and the idea that this one and a half percent difference - this difference gained in the tiny evolutionary period since we split from our common ancestor - can fundamentally rewrite the rules of evolved behaviour is just plain wrong.

But the heresy is even worse than that. As Diamond goes on to explain, although we split from a common ancestor with the chimpanzees some six plus million years ago "[f]or most of the time since then, we have remained little more than glorified chimpanzees in the ways we have made our living". Our ancestors remained "little more than glorified chimpanzees" - with all that that should imply - until a couple of hundred thousand years ago. And these glorified chimpanzees were genetically even closer to us, until, in the last instance, "[p]erhaps they shared 99.9% of their genes with us". There is no evidence whatsoever that these "glorified chimpanzees", *Ardipithecus*, perhaps several distinct species of *Australopithecus*, *Homo habilis*, *Homo erectus*, and the only slightly more advanced archaic *Homo sapiens*, managed to break from the pattern of gross immorality that runs throughout the natural world. These ancestors, finally only maybe 0.1% away from modern man in their genetic makeup (such a tiny variance that this same 0.1% is the genetic distance between two modern human beings selected at random), were still coded for cannibalism and infanticide. And evolutionary psychologists accept this, which is why they postulate that morality emerged from gross immorality some

---

includes indels, insertions and deletions of DNA found in one species but not the other. There is some dispute about the significance of Britten's finding, partly because indels are common in non-functional sections of DNA but it is their occurrence in genes themselves that would really matter. However, even if we were to find that indels have an important effect here, since indels exist across other species boundaries too it almost certainly does not change the observation that humans are more closely related to both species of *Pan* than both species of *Pan* are to the other apes.

100,000 years ago in that last genetic blip which produced modern *Homo sapiens*. Yet (even ignoring the non-existence of such a selecting mechanism) it is an evolutionary *impossibility* that we could have retained 99.9% of a chimp's DNA (glorified or otherwise) and expect that final infinitesimal 0.1% to have been the focus of the drive for selfishness as old as life itself. Gross immorality is a four billion-year programme, not a chance susceptibility to malaria. No, for Diamond, as for almost all biologists who have thought about this question, something *extra* was added in that last 0.1% change. We didn't *lose* anything (and certainly not a four billion year pattern for extreme selfishness), we *gained* something. We gained a susceptibility to culture with the emergence of the capacity for spoken complex language (although this susceptibility also required a very large brain which had itself taken millions of years to evolve. In chapter 10 we consider the importance of our highly evolved capacity to mentally simulate and note that simulation seems to exist, but at a vastly reduced level, in social animals such as chimpanzees). Evolutionary geneticists have long known that this tiny change could have easily added the necessary genes, and that both speech and brain size may depend on very few new genes. As *New Scientist* noted on 15 May 1999 ("The Greatest Apes", Karen Hopkin) human brains are twice as big as chimp brains, and our two brain hemispheres have different connection densities from chimps - "A single gene could lie behind either of those differences". The scientifically-reasonable view of the leading selfish gene theorists (and many other Darwinians) is that we received something extra in that last 0.1% shift. Just compare this view with the evolutionary psychology alternative that nature backtracked through the human genome, deleting or inhibiting genes for the great majority of animal behaviours while substituting genes for a vast number of new human behaviours.

Mutations with large effects, or macromutations, do occur, and where they produce a radical difference in the visible characteristics shown by an organism such is termed a "monster". But macromutations cannot contribute to evolution because they will be eliminated by natural selection. The great majority of mutations are deleterious to offspring, and are consequently removed by natural selection. Existing organisms are well adapted to their environments, and any random mutation is a game of Russian roulette with a very small likelihood of an empty chamber. While the aftermath of a mutation is dependent on the gene affected and the nature of the mutation, natural selection's game of tinkering with genetic material can be not only ultimately hazardous, but immediately lethal. There are perhaps 30,000 genes in the human genome, arranged along 23 pairs of chromosomes. There are a total of 3 billion base pairs of nucleotides in a human being (the basic units of nucleic acid molecules found in the genes), and while only a small percentage of these are functional this still equates to more than 100 million coding base pairs. Yet a mutation that causes the loss of just 3 nucleotides on a single gene on chromosome 7 is the most common cause of cystic fibrosis, an illness where thick mucus builds up in the lungs and causes life-threatening infections. While today much can be done for patients with cystic fibrosis to help them now live beyond childhood, and much more may potentially be done with future gene therapy, the point is that in nature such a tiny mutation would cause swift death in youngsters. So even tiny mutations are a risky business in nature, and the more mutations that occur, the higher the risk of problems.

But occasionally a mutation does occur which is not damaging to offspring and also gives an actual advantage over others. Even a beneficial mutation will probably be lost by chance, but, given the frequency of small mutations within large populations, such a beneficial mutation will reappear and will sooner or later begin to be favoured by selection and ultimately incorporated into the gene pool of the species. But, as should now be obvious, modern biology rejects the notion that nature can work through huge leaps in organism design, that you can throw away much of what goes before. As Dawkins puts it, "The greater the number of simultaneous improvements we consider, the more improbable is their simultaneous occurrence". An associated twist is that most complex characteristics under genetic control are, and have to be, polygenic - that is, governed by the combined interaction of a number of genes. If you argue that in that last genetic spurt man inherited genetic coding for all sorts of new moral attributes you are positing that chance addition of multiple new (working) gene complexes. To return to the *New Scientist* point, changing brain size several times since departing from that common ancestor with the chimp, or modifying connection densities - here you are only adding in stages to what already exists, just tinkering with what is already there in a very straightforward (albeit time-consuming) way. In contrast, evolutionary psychology's theories would require entirely new fully wired-up behavioural mechanisms, even the foundations of which could not conceivably have existed in our ancestors. We are thus into the realm of astronomical statistical improbability. So to assume that our ancestors somehow broke from the billion-year pattern of nature, losing the natural world code for gross immorality and gaining the code for morality and virtue, is simply not realistic. Even if you posit not one gigantic leap but two or three still giant leaps you face this same problem of statistical improbability, because at some stage you are requiring that one genome-wide behavioural code (gross immorality) was simply deleted or abandoned, and that its opposite (morality) was put in its place.

Nature does not make leaps. She did not re-write the great majority of our behavioural code in one huge genetic bound 100,000 years ago, turning a grossly immoral ape into a moral human, as claimed by human sociobiologists. Neither did she evolve an anti-Darwinian "conscience module" that then inhibits the expression of 99.9% of the behavioural genotype of an otherwise grossly immoral ape, as effectively claimed by some evolutionary psychologists[12]. The idea that nature can act through vast beneficial macromutations is the *bête noire* of modern evolutionary theory; it is utter heresy. The hypothetical products of such fanciful speculation are known derisively as "hopeful monsters", after Richard Goldschmidt's 1930s work which suggested that saltation led to speciation. Such as, presumably, the leap that turns a grossly immoral ape into a moral human.

And what a leap. The complete re-writing of a four billion year programme, a programme written across 3 billion nucleotides. Not just another selfish ape, set free by culture acting on a very large brain, but an ape with an almost completely re-written behavioural genetic code. An ape that had, in that last genetic spurt 100,000 years ago, lost its species-wide coding for cannibalism, for infanticide, for extreme violence, for pitiless indifference, for gross immorality. Say hello to "the virtuous ape". But this is not Darwinism; this is anti-Darwinism. As the

---

[12] Evolutionary psychologists are wary about making this claim explicit, because the idea that genes can so easily be inhibited gives them the willies. However, as we have seen, their only other option is to claim that our grossly immoral genes just happen to have a fortuitously unique "maladaptive" phenotypic expression in today's novel surroundings (see Donald Symons' hypothesis earlier).

Natura non facit saltum

Darwinian philosopher Daniel Dennett notes: "People ache to believe that we human beings are vastly different from all other species - and they are right! We are different. We are the only species that has an *extra* design preservation and design communication: culture".

Life and all its glories are thus united under a single perspective, but some people find this vision hateful, barren, odious. They want to cry out against it, and above all, they want to be magnificent exceptions to it.

Daniel Dennett – Darwin's Dangerous Idea

In the last few decades, the only people *incapable* of accepting that genes influence human behaviour have been human sociobiologists, evolutionary psychologists and behavioural geneticists. Genetically, we are just another selfish ape. As Darwin realised, it is culture that makes us human.

But why are the evolutionary heretics so desperate to deny that humans are products of orthodox natural selection? Why does the idea that we may have the same behavioural genetic code as all other primates so frighten them? Just what do we mean by the "gross immorality" of nature?

4

A world without compassion

Simple cannibalism ... can be expected in all animals except strict vegetarians.
George C. **Williams** – "Huxley's Evolution and Ethics"

What is our genetic code then, if we strip away the anti-Darwinism of the last twenty-five years and accept what Williams, Maynard Smith and Dawkins are telling us? If we accept that we never broke the billion-year mould, that we never evolved to be the *decent*, the *virtuous*, the *moral* ape?

Human decency is not "animal", as Edward Wilson first claimed. Decency does not, *cannot*, exist in nature. In the nineteenth century, Darwin's Bulldog spoke of the moral indifference of nature. Culture, said Huxley, combats biology, and in so doing makes civilisation possible. Today biologists go much further. Dawkins speaks often of the "blind, pitiless indifference" of nature. Over a decade ago, George Williams rejected Huxley's term moral indifference and substituted his

A world without compassion

own term the "gross immorality" of nature, our "enemy". But why did Huxley understate his case? Just what is it that selfish gene biologists have discovered about nature?

As Stephen Jay Gould tells us in his essay "Ten Thousand Acts of Kindness", "*Homo sapiens* is a remarkably genial species" (reprinted in *Eight Little Piggies*). Now, because Gould belonged to the hierarchical, or multilevel selection, tradition which sees a role for group-level selection, he could put such behaviour down to genetic group selection; the idea that animals can be born programmed to act for something other than their own selfish genetic interests. The scale of human kindness still shocked him, but as the hierarchical school wishes to get away from the selfish gene tradition and its stressing of the relentless selfishness of nature, Gould didn't realise quite *how* shocking it is. But selfish gene theorists do not have this outlet, as to invoke genetic group selection to explain major behaviours is heresy. And because the evolutionary psychologists largely deny that behaviour can be the result of anything other than genetics, they can only cling to "implicit" group selection. Human sociobiologists and evolutionary psychologists do not deny the uniquely moral behaviour in man (the standard line found in all their books is that "human morality is a biological adaptation"), they just step outside orthodox Darwinism to explain it.

Before the advent of selfish gene-ery, when instances of cannibalism and infanticide were noted in natural world studies they were dismissed as aberrations. The field workers could not understand such behaviour, and so they chose not to concentrate on it. All this began to change in the 1970s as field work became more common and rigorous, researchers began to share and discuss findings more widely, and, most importantly of all, Darwinism was interpreted through a refinement that understood the adaptive significance of what was being observed. "The story of the forest or coral reef is a tale of relentless arms races, misery, and slaughter", writes George Williams in *Plan and Purpose in Nature*. This in itself is not a new understanding; nature "red in tooth and claw" is an insight that pre-dates Darwin. The *scale* of the horror is still shocking to modern evolutionary theorists of course, a scale not even imagined in Darwin's day. Williams has noted that other wild animal populations are many *thousands* of times more likely to kill than are humans from even the most murderous of American cities. Stephen Jay Gould has the same point to make in his essay "Ten Thousand Acts of

A world without compassion

Kindness". Ethologists, he notes, describe organisms as peaceful if tens of hours go past between aggressive encounters. Consider the thousands upon thousands of hours that can go past for human groups, he says. The primatologist Frans de Waal notes likewise (italics his): "What is most amazing is that our species is able to survive in cities at all, and how relatively *rare* violence is". De Waal highlighted the eloquent way in which the American ethologist Nicholas Thompson drew attention to the contradiction in 1976: "[W]hat is surprising about people in subways is not their hostility; on the contrary, it is the degree of coordination and habituation which permits thousands of people to move daily through an environment so physically hostile as to stampede the herds of any sane animal." But what was distinctly new was not so much the awareness of the scale of the violence and horror in nature, as the awareness of the utter immorality of nature. Chimpanzees, admits de Waal, live in "a world without compassion". Gone is the bright optimism of Darwin's day that nature might still teach moral lessons to man, as Williams explains. "With what other than condemnation is a person with any moral sense supposed to respond to a system in which the ultimate purpose in life is to be better than your neighbor at getting genes into future generations, … in which that message is always 'exploit your environment, including your friends and relatives, so as to maximize our (genes') success', in which the closest thing to a golden rule is 'don't cheat, unless it is likely to provide a net benefit'?"

Williams continues, "[a]wareness of the prevalent wickedness of what had been personified as Mother Nature is a recent development". Now naturalists find evidence of gross immorality wherever they look; the cannibalism, the killing of others' young, the killing of one's own young, the rape, the indifference to suffering. "Mountains of data on parasitism and predation (including cannibalism) in nature could be amassed to document the enormity of the pain and mayhem that arise from adaptations produced by natural selection", says Williams. In his academic work Williams has been even more graphic, and much more detailed. In his 1988 *Zygon* paper "Huxley's *Evolution and Ethics* in Sociobiological Perspective" he devotes over seven pages simply to detailing the mountains of data already collected by field researchers. "Besides adultery and rape, just about every other kind of sexual behavior that has been regarded as sinful can be found abundantly in nature. Brother-sister matings are the rule in many species (Hamilton

A world without compassion

1967). ... Homosexual behavior is common in a wide variety of vertebrates (Beach 1978), and perhaps homosexual preference (Weinrich 1980)." The following year Williams expanded his paper (and his selection of sources) into a book co-authored with the historian of science James Paradis. "Simple cannibalism is the commonest form of killing, and Polis' (1981) review indicates that it can be expected in all animals except strict vegetarians".

You see, from the viewpoint of the selfish gene this is all supremely logical; all the rape, all the infanticide, the cannibalism and the extreme violence. Darwinism is about *efficiency*, it is about maximising inclusive fitness, and any behaviour that aids the transmission of genes into the next generation - no matter how grossly immoral from a human perspective - should be expected to be found abundantly in nature. There is no grief in nature, because grief is a waste of resources. You do not grieve over a dead infant, you just get on with the job of having another. Individuals are what Dawkins first called "gene vehicles", and nature has designed parents to maximise the number of genes getting into the next generation through these vehicles. If you have X young but the weather takes a turn for the worst, you potentially damage your chances of maximising surviving genes if you try to raise them all. Why therefore raise X weak infants when a slight logistical change will give you X-1 very healthy gene vehicles? Identify the weakest of your young (usually the last to be born), and stop feeding it. Better still, tear it up as food for the others (it might be small, but it is still an easily available source of protein and valuable nutrients). No one else in the nest or den is complaining. What once they called brother, now they call lunch. "There is no charity in nature," the geneticist Steve Jones reminds us. Not even a charity that begins at home.

Rape was a strategy that was widely expected by selfish gene theorists even before the observations began flooding in. In one of the earliest detailed papers on avian rape, Pierre Mineau and Fred Cooke noted in 1979 that even the briefest absences by her mate could make the female snow goose vulnerable to rape by neighbouring males. Rape makes sense when the only thing that drives life is the need to get your sperm to fertilise as many eggs as possible. Nature has made males cautious about leaving their mates unguarded for long, because in addition to rape he runs the risk that she will seek a finer gene pool than is offered by his seed. Two can play the game of enhancing gene survival.

A world without compassion

There is no compassion in nature, no consideration for victims, no pity, no regrets. There is no love either; some animals (such as albatrosses) may mate for life, but all the evidence is that this is through necessity, not choice. Animals that partner for life have found that this is the best way of ensuring the long term success of their offspring; their commitment is to their genes, not to each other, as is demonstrated if they get an opportunity to cheat on the other. There is no true love for offspring either. Natural selection is about *sacrifice* for those that carry your genes, not what we would understand as *love*. A mother may fight savagely to protect her offspring, but when she knows the game is lost her behaviour changes immediately. Weak offspring are killed if this is the most efficient strategy. Mother chimpanzees mate contentedly with their infant's killers. And no grief, of course, since one gene vehicle is likely to be as good as another, and just as much a biological tool. Smaller mammals, with less access to reliable sources of food, and worn down by a long period of having to suckle, will kill and eat their own offspring rather than allow them to fall into a predator's hands. This is because a healthy mother will get a chance to mate again next season, so it is time to build herself up rather than waste energy fighting the inevitable. Why risk serious harm to yourself when the product is replaceable? Infants are important, sure, but you can have others next season, so they are only worth fighting for up to a point. The point when the costs outweigh the benefits.

As with the vices, so, inversely, with the virtues. While vice is found in abundance in nature, virtue is unknown. Duty, compassion, charity. All unknown in nature because nature is about individual selfishness, not the well-being of groups.

"An extraordinarily murderous lot"
When a lion acquires a new lioness who is still nursing cubs from an earlier mating, the first order of business is to kill those cubs, so that the lioness will more quickly come into estrus. ... [L]angur-monkey males often kill the infants of other males to gain reproductive access to females. ... This dark message about our furry friends is often resisted.
Daniel Dennett – Darwin's Dangerous Idea

Williams, in *Plan & Purpose*, notes that it was the anthropologist Sarah Blaffer Hrdy who was the pioneer in bringing the prevalence of monkey infanticide to the attention of both biologists and the general public. Her 1977 article "Infanticide as a Primate Reproductive Strategy" was greeted "with outraged disbelief by many readers who refused to believe that adult males' attacks on infants could be adaptive and normal". Times have changed, he notes. He recounts the story by the psychologist Martin Daly of a 1993 conference where a biologist was relating how a species of carrion beetle uses the corpse of a dead mouse as a food source for her newly-hatched young. "A challenger may sometimes appropriate a dead mouse from a guarding female. Someone in the audience asked, at the close of the presentation, about what happens to the young of the first female. The speaker immediately explained that the new female kills them, 'of course'. Daly regretted Hrdy's absence, imagining that she would have felt a smug satisfaction at realizing that her heretical findings were, less than twenty years later, simply expected."

(We shall meet Daly again in a later chapter. He developed Hrdy's findings - fully detailed in her book *The Langurs of Abu* - and subsequent human speculations to suggest that where infanticide occurs in human societies there must also be a genetic basis for it. Daly has missed the point that when Hrdy was documenting the casual slaughter she witnessed in langur monkeys, she admits that it was her own tears that were among the problems she had to overcome. While *every* adult langur or chimpanzee will tear infants of their own species to pieces without batting an eyelid, humans shed tears over the deaths of infants of *another* species. Humans, the uniquely "decent ape"[13]. Daly is an evolutionary psychologist; evolutionary psychologists believe that humans evolved a moral capacity - it's just that some of us (through different upbringings) have less of this anti-Darwinian morality than others. Evolutionary psychology can be contrasted with human sociobiology; human sociobiologists - and behavioural geneticists - tend

---

[13] In his classic 1966 work George Williams provides us with a wonderful example of the utter incongruity between human sentimentality and natural world pitiless indifference. He recounts the attitudes of an audience being shown a film about elephant seals: "Amid the crowded but thriving family groups there was an occasional isolated pup, whose mother had deserted or been killed. These motherless young were manifestly starving and in acute distress. The human audience reacted with horror to the way these unfortunates were rejected by the hundreds of possible foster mothers all around them".

to just think that some of us are born with less of this anti-Darwinian morality. Most of us are fortunately born human; some of us are unfortunately born more ape than human. In chapter 7 we shall consider whether those rare instances of human infanticide actually have any residual genetic component at all.)

It is rather difficult to separate cannibalism and infanticide in nature, because most slaughter is of the young. Cannibalism for food does occur, and obviously the smaller it is the less of a risk it will put up a fight. The rule of nature is - "if it's smaller than you, prey on it; if it's bigger than you, run from it". However, cannibalism is often linked to more pressing relationship issues, and the need to clear out the genetic opposition, though the necessity of the kill (and of quickly getting back to your young) overrides the nutrition element. Williams notes the differential infanticide strategies nature has developed in ground squirrels. "A male may raid a nest to kill and eat one of the young. A female may raid the nest of a competitor and kill all the young (but not eat them)", (in Paradis and Williams). While it is always more difficult for a male to keep track of which offspring he has fathered (even usually in species where adults pair for life), it is much simpler for the female to identify the threat posed to the hegemony of her genes and do something about it. Yet when males are certain of their lack of relatedness, they may indulge in more spectacular cannibalism, as we will see below.

Such infanticidal slaughter exists throughout the mammalian (including primate) worlds. Hrdy's groundbreaking paper on monkey infanticide argued that infanticide by males could no longer be dismissed as "abnormal". Her own observations on langur monkeys at Abu in India allied to the work of others "led me to reject my initial crowding hypothesis [*as a cause of infanticide*] in favor of the theory that infanticide is adaptive behavior, extremely advantageous for the

---

Williams' point was to show the fallacy of "good-of-the-species" theorising - "It should have been abundantly clear to everyone present that the seals were designed to reproduce themselves, not their species". However, the passage serves equally well to demonstrate the leap of faith (not to mention rejection of over one hundred years of evolutionary theory) it takes to even begin to believe that such a grossly immoral process as natural selection could produce common human decency. Either audience reactions of horror or the tears of Sarah Hrdy.

males who succeed at it." Infanticide was "the single greatest source" of the up to 83 percent infant mortality rate she found at Abu. Equally shocking to her were the instances of mothers abandoning their murdered infants soon after or even before death. She considers whether, as some had suggested, this was caused by a mother's fear, but rejects it in favour of a more hard-nosed gene-selectionist understanding: "It is far more likely, however, that desertion reflects a practical evaluation of what *this* infant's chances are weighed against the probability that her next infant will survive".

Cannibalism itself is not an homogeneous act, and zoologists must make a distinction between heterocannibalism (of unrelated individuals), and sibling and filial, or parental, cannibalism. All of these forms of cannibalism can occur in different species, and have a range of alternative explanations. Cannibalism, note Elgar and Crespi (1992), occurs in a variety of contexts, including infanticide, mating and courtship, adult-adult cannibalism, and competitive interactions. One obvious feature is that cannibalism tends to be associated with an asymmetry between cannibal and victim. Generally the victim is at a more vulnerable stage of its life cycle, so while cannibalism between adults does occasionally occur, cannibalism of juveniles by other juveniles is more common, and cannibalism of juveniles by adults can be very common. Asymmetrical cannibalism has evolved because selection has favoured it where rewards (in terms of nutrition, reproductive advantage, sexual advantage or competitive advantage) exceed costs (in terms of conflict or potential loss of fitness).

Cannibalism is as common in marine vertebrates as it is in birds and dry-land vertebrates. More than 160 years ago geologists were finding evidence of cannibalism in the petrified faeces of 200 million-year-old marine reptiles (Buckland [1835]), and today it is marine cannibalism that provides some of the most stunning examples of the benefits of cannibalism exceeding the costs. In 1948 intrauterine (or within-the-womb) cannibalism was discovered in the sand tiger shark (*Carcharias taurus*, also known as the grey nurse shark), where the largest and strongest embryos eat their weaker siblings in the process of adelphophagy (literally "consuming one's brother"). Sand tigers have two oviducts, and the first young in each oviduct to reach 6 cm swims to the uterus where it feeds on its siblings; a large part of the reason that it is commonly around 100 cm at birth. The major benefits from an

A world without compassion

evolutionary point of view are that surviving sand tiger young experience very rapid growth, are very active predators at birth, and often have a considerable size advantage over other predators. The pike-like freshwater game-fish the walleye has provided examples of cannibals within cannibals within cannibals, as larger walleyes were found to have eaten smaller walleyes, which had eaten still smaller walleyes, for at least a four-fold cycle. Cannibalism in marine mammals is also now well documented. Cannibalism has been seen among grey seals in Canada and elephant seals in Argentina, but it was only as recently as 1999 and 2000 that cannibalism was first recorded in sea lions, when 24 cases were seen in just 12 weeks in one colony of New Zealand sea lions.

Hrdy's 1977 paper also quoted others' recent findings; from Dian Fossey's observations of infanticide in African gorillas to David Bygott's graphic recounting of cannibalism in Tanzanian chimps. "The female and her infant were immediately and intensely attacked by the males. For a few moments, the screaming mass of chimps disappeared from Bygott's view". Hrdy recorded that when Bygott managed to relocate them, the strange female had disappeared. "One of the males held a struggling infant. 'Its nose was bleeding as though from a blow, and [the male], holding the infant's legs, intermittently beat its head against a branch. After 3 minutes, he began to eat the flesh from the thighs of the infant which stopped struggling and calling'". In contrast with normal chimp predation, this cannibalised corpse was nibbled by several males but never actually consumed, Hrdy wrote. Bygott's 1972 work was one of the very first to catalogue chimpanzee cannibalism, and it detailed how even low-ranking group members indulged in such sport: one chimp "(formerly the dominant male of the community but now low ranking), approached and was permitted to tear off one of the infant's feet". Only a decade after Bygott was writing, the geneticist Steve Jones would report that cannibalism had by then been recorded in more than 1,300 species of animal and was often the primary cause of mortality. Hrdy finished her survey by concluding that: "we are discovering that the gentle souls we claim as our near relatives in the animal world are by and large an extraordinarily murderous lot".

Let us finish, then, on chimpanzees. *All* chimpanzees will tear apart the male infant of one of their females where the infant may have been fathered by a chimpanzee from another group. The reason? Simply because the infant is a potential genetic rival. The group will not waste valuable resources protecting this new member when it may bear no genetic relationship to any of the males of the group. The mother, too, is a resource too valuable to waste (a new womb for the males), and she becomes a willing party now in realising that latent value. Within a few days she will be found happily locked onto the loins of one (and often more) of her infant's killers, ready to conceive a replacement gene vehicle. As Williams notes of a similar group of primates: "[t]he death of her infant converts her more quickly from a potential to an actual resource for the male's reproduction. This is why infanticide is adaptive for the male." But selfish gene theory makes females no more dainty than males. Cannibalism in female non-human primates does occur, albeit less frequently, as both initiators and sharers of the spoils. Within chimpanzees, adolescents and juveniles of both sexes will sometimes share cannibalised meat. The primatologist Mariko Hiraiwa-Hasegawa, who puts the frequency of cannibalism in chimpanzees as "exceptionally high", notes that a female chimpanzee and her adolescent daughter initiated three separate acts of cannibalism as witnessed by Jane Goodall and colleagues in Gombe National Park between 1975 and 1976.

Meanwhile, Lukaja *handed the infant to the alpha male* Ntologi, *who dragged, tossed, and slapped it against the ground.* Ntologi *climbed a tree with the infant in his mouth. He waived it in the air, and finally killed it by biting it on the face. ... Conspicuous competition for meat and meat-sharing was observed as usual. Three adult males and an adult female obtained meat from* Ntologi. *Two adult females, two juvenile females, a juvenile male, and an infant recovered scraps from the ground or were given scraps. At 13:00,* Ntologi *was still holding the skin of the carcass.*

Hamai et al. – "New records of within-group infanticide and cannibalism in wild chimpanzees"

Bonobos: A new kind of ape?

"De Waal ... is distressed at what he mistakenly sees as a neo-Darwinian tendency to emphasise the 'nastiness of our apish past'. Some of those who share his romantic fancy have recently become fond of the pygmy chimpanzee or bonobo as a yet more benign role model" (Dawkins, *Unweaving the Rainbow*). As we have seen, genetically bonobos are as closely related to humans as are chimpanzees, having split from each other a million or so years after mankind's common ancestor split from the common ancestor to both bonobo and chimp. The fact that chimps are ecologically more diverse than the bonobo might suggest that bonobos, and not chimpanzees, are the derived species, but there is as yet no real hard evidence to prove this.

Importantly for this discussion, infanticide (and, it almost goes without saying, cannibalism) has not yet been witnessed in bonobos. Certain primatologists like to draw attention to the relatively peaceful nature of bonobos, that they "make love, not war". Some suggest that bonobos might be better models for early man (if one can overlook the fact that bonobos regularly indulge in sex in every possible combination, including male with male and adult with young juvenile, and have no exclusive sexual orientation). There are some facts about bonobos that should, however, be considered.

Craig Stanford used the pages of *Current Anthropology* in 1998 to point out that the number of field observation hours on bonobos is still a small fraction of the hours spent field observing chimpanzees. Stanford also noted that half of all encounters between bonobo communities do still result in aggression of some sort, and that observer bias, including the overemphasis on captive, rather than field, observation of bonobos, may be misleading the public and scientists alike. While infanticide and lethal intercommunity aggression are both "common" in chimps, they are still restricted to rare periods of time. Understandably natural selection has made such instances ephemeral, simply because they are so predictable. Natural selection will tend to come up with a counterstrategy where possible; if males are going to kill unrelated infants, females will tend to have been selected to steer clear of males while infanticide can advance oestrus. Similarly, intercommunity conflicts are predictable but tend to be short-lived. Consequently chimpanzees had to be observed for more than 15 years before lethal intercommunity aggression was finally witnessed.

Many of Stanford's points were well taken by other bonobo researchers, especially the field researchers. Katharine Milton reminded her colleagues that: "we do not have to look far into the past to recall how, as more field data emerged, the sunny image of the playful, fruit-eating chimpanzee at Gombe was gradually revealed to have a darker side". Takayoshi Kano, leader of the bonobo research project at Wamba, noted that lethal bonobo intercommunity aggression may indeed one day be witnessed, citing his own observation of "a severe laceration" on one young adult that got separated from his main party for a few days. And de Waal was quick to point out that he himself has said that infanticide may yet be witnessed in bonobos when more study has been undertaken. Such hedging of bets seems wise. Noting that infanticide by males is most advantageous where lactation is long relative to gestation, van Noordwijk and van Schaik (2000) predict that all great apes are vulnerable to infanticide, even though, whilst well recorded in gorillas and chimpanzees, infanticide has not yet been witnessed in orang-utans or bonobos. Frequency of expected attacks, however, is partly a function of the counterstrategies females have evolved. Female orang-utans are semi-solitary. Thus while female orang-utans are exposed to male attacks (including documented rape) whilst actively receptive to sexual encounters, after giving birth mothers with infants rarely associate with conspecifics. Frequency of infanticidal attacks is thus expected to be low but not zero.

Bonobos: Old wine in new bottles
To argue that bonobos are relatively peaceful, and perhaps therefore even "proto-moral", is entirely to miss the point grasped by selfish gene theorists. There is no morality in nature. Like orang-utans, the reason that bonobos rarely, if ever, commit infanticide is purely because they do not get the opportunity. Bonobo groups are female-dominated, whereas chimpanzees are male-dominated. In bonobos, females get to set the rules. One rule they set was sex. De Waal (in *Bonobo: The Forgotten Ape*) comments that what the female bonobos are doing might be construed as an anti-infanticidal strategy. "They have managed to make paternity so ambiguous that there is little to fear. Bonobo males have no way of knowing which offspring are theirs and which not." Male chimps, de Waal notes, have a "tendency" to kill identified unrelated infants, so it is "no wonder" that chimp females stay away from large gatherings of their species "for years after having given birth". No such worries for female bonobos, however, because a male bonobo who knew his offspring "would have to be a genius": "If one had to design a social system in which fatherhood remained obscure, one could scarcely do a better job than nature did with bonobo society". Furthermore, notes de Waal, since bonobo females tend to be dominant, attacking them or their offspring would be a risky business. Most likely if a male were to be

perceived as a threat to any infant females would band together in defence.

De Waal has written that chimpanzees live in "a world without compassion". Yet bonobos, too, live in this gene-selectionist world without compassion, and bonobo behaviour can be every bit as vicious as chimpanzee behaviour. Primatologists have been intrigued by the high rate of physical abnormalities in the male bonobos at Wamba. As Amy Parish told de Waal in interview, this could have something to do with the females. Females in captivity establish dominance over male bonobos by overt aggression. At one zoo, noted Parish, the females occasionally held down the male and attacked him, and had bitten off parts of his fingers and toes. At another zoo the alpha female had a similar relationship with her adult male: "It is assumed that she once bit his penis almost in half", Parish reported.

In bonobo groups the whole strategy of infanticide by males becomes counterproductive, admits de Waal, because males can never be sure they are not killing their own offspring. Females have managed to make paternity so ambiguous that "there is little to fear". But anyone who wishes to argue that such constrained behaviour demonstrates proto-morality or a "good nature" would have to conclude that a man-eating shark in a secure aquarium is similarly "good natured" simply because it had not yet found an opportunity to devour its keeper. There is a reason why bonobos do not co-exist in groups larger than a few dozen, and that reason is the gross immorality of the genetic world.

Our nearest relatives: "an extraordinarily murderous lot". Cannibalism, infanticide, rape, and death from strife rates not far off one *million* times higher for wild animal populations than the homicide rate of a small rural Western European population. Cannibalism, infanticide, extreme violence, "gross immorality". The programme of the selfish gene. The programme of natural selection. The genetic programme of *every* human being that has ever lived.

The Book of Man
Within animals we are vertebrates, within vertebrates we are mammals, within mammals we are primates, and within primates we are - by exactly the same logic - apes.
       Richard Dawkins – "Sociobiology: the New Storm in a Teacup"

A world without compassion

Yet the behaviours investigated above are the programmes for all non-human animals. These various behaviours exist across the natural world because such behaviour *is* the natural world. You will find it for all animals that exist in the wild, including those we tend to find cute or noble, even the "gentle souls", as Hrdy put it. Apes, penguins, dolphins, and even domestic dogs if you wait until they get hungry enough[14]. This is the behaviour that billions of years of evolution have discovered *works*. This is the selfish gene, in all its glory.

This lack of compassion, this gross immorality, exists for all animals, and for all members of a species. Such behaviours are not *by-products* of naturally selected behaviour; they are the *dynamics* of natural selection. An animal that is busy protecting its mate from rape this season will be busy raping others' mates next season. As Mineau and Cooke noted of their snow geese, personal status, familiarity and coupling matter little in the game of gene preservation. This is the universal genetic logic of natural selection where only opportunity matters: "The rapist also was usually a known territory holder (84%), commonly a neighbour. Rapists seem to capitalize on attendant male absenteeism... An absent male is himself usually (73%) raping another female or witnessing another rape. Indeed, 'gang rapes' usually occur when spectators at a rape attempt [to] use the disturbance to join in the melee". And there are not some male primates out there who will rape, and others who will not. There are not some male and female chimpanzees that will tear apart the male infant of a mother who has recently joined their group, and some who will not. Darwinism tells us about the universal, and grossly immoral, nature of each species.

In contrast to the anti-Darwinian propaganda of the human sociobiologists, the behavioural geneticists and the evolutionary psychologists, the "correct" genetic code for man is not to be a middle

---

[14] Domestication is a thin veneer. Even untold generations of intense artificial selection in the descendants of the wolf have been able to barely mask the gross immorality of the natural world. As Steve Jones notes in *Almost Like A Whale*: "By owning a dog, any dog, men welcome into the home a beast that preserves much of its primordial self. Overgrown juveniles though they are, evolution by human choice has not removed the instincts of their ancestors. ... Like wolves, dogs attack the weak, be they children, old, or drunk. Packs of feral animals have pulled infants from bicycles and eaten them, and a mere half-dozen beagles, dachshunds and terriers once devoured an eighty-year-old woman."

class conservative. It is to be an ape; it is to be just another selfish ape. For the reasons we saw in the last chapter, natural selection did not break the mould when it came to man. Nature does not, *cannot*, programme in any other way. **Mankind has a universal, *innate* nature. It is a nature shared with every other *ape*. A nature shared with every other *mammal*. It is the nature of the selfish gene, and of pure, unalloyed, genetic self-interest. It is a nature of sexual eclecticism, horrific violence, cannibalism and infanticide.** An innate nature shared by every man, woman and child that has ever walked on this planet. Elizabeth Windsor, the Queen of England, was born a cannibal. Karol Wojtyla, 263rd Pope and Vicar of Christ, is genetically programmed to tear the arms off a human infant faster than a small child can pull the limbs off a daddy-long-legs.

The next time someone tries to tell you about a hypothetical gene for violence carried by only a few people, or by only one group of people, patiently explain to them about the gene for cannibalism carried by the entire human population. Explain that they, like you, carry genes that would make them pull babies to pieces and feel absolutely no remorse. Genes that we know exist, because they exist in every chimpanzee. Explain how genes cannot code for qualities such as love, grief, duty, honour or virtue. Explain how none of the things that make us truly human - neither love and honour, nor hate and racism - can come from any gene pool.

Contrary to the anti-evolutionary propaganda of the last twenty-five years, we are products of the natural world. And Darwin's heinous crime, for evolutionary psychologists like Helena Cronin, was simply to suggest that we carry the same "extraordinarily murderous" biological code as all other primates.

From the brutes, but not of them

At the same time no one is more strongly convinced than I am of the vastness of the gulf between civilized man and the brutes; or is more certain that whether **from** *them or not, he is assuredly not* **of** *them.*

Thomas Huxley – Man's Place in Nature

Even in the nineteenth century, biologists knew that we are not *of* the "brutes" simply from basic behavioural comparisons. Today, when we know of the gross immorality of the natural world, the difference is more immediately obvious. Queen Elizabeth II is not a cannibal and Pope John Paul II is not an infanticide, despite continuing to carry the genetic codes for cannibalism and infanticide. While some accuse the current Queen of being distant from her people, no one has yet accused her of eating them. She is *from* the brutes, but she is not *of* them. So what can be happening? Man takes no part in the natural world species-wide practice of cannibalism, of infanticide, of gross immorality. Yet we know that the attempt of the evolutionary psychologists to escape nature by positing genetic uniqueness is false. What is taking place? There can be only one answer, of course. "For a geneticist, all variance which is not genetic is, by definition, 'environmental'", writes John Maynard Smith. Queen Elizabeth II is not a cannibal and Pope John Paul II is not an infanticide because culture got in the way. Like each and every one of us, Elizabeth Windsor and Karol Wojtyla were born ape, but made human.

Culture is not the product of biology, the claim of the evolutionary psychologists. Culture must *combat* biology, as Thomas Huxley said. And as Williams, Maynard Smith and Dawkins have all written. This is the reason that Dawkins needs his "memes" to fill the gap, and why Williams and Maynard Smith write about the tremendous power of culture. This is why Darwin went looking for the factors in human culture that were powerful enough to overcome our murderous biology and allow us to act virtuously, factors that he found in positive and negative peer pressure, religion and ritual. As Darwin wrote in *The Descent of Man*: "a belief constantly inculcated during the early years of life, whilst the brain is impressible, appears to acquire almost the nature of an instinct". Factors that could take an ape, and make a human.

The former British Prime Minister Margaret Thatcher, echoing the egoist philosopher Ayn Rand, once said that: "There is no such thing as society. There are individual men and women, and there are families". But Darwinism tells us that the *only* thing is society, because without society there would be no human individuals, and no human families. Only apes. This is simply a process of elimination: what *cannot* be genetic *must* be cultural. But simply knowing that culture must be allowing us to rise above genetic control does not explain how and why culture makes this possible. This is Huxley's "apparent paradox": "that ethical nature, while born of cosmic nature, is necessarily at enmity with its parent". Huxley himself thought the answer to resolving that paradox lay in our capacity for reason. "In virtue of his intelligence, the dwarf bends the Titan to his will", he wrote. "Fragile reed as he may be, man, as Pascal says, is a thinking reed: there lies within him a fund of energy, operating intelligently ... that is competent to influence and modify the cosmic process." This was also to be the answer proposed by the philosopher Peter Singer in 1981 in *The Expanding Circle*, his critique of human sociobiology. Yet Maynard Smith, for one, remains to be convinced: "The difficulty for any account of society that assumes that individuals behave rationally is partly that experimental psychologists find little support for such an optimistic view". Gould, in typical style, is even more forthright: "Before Freud, we imagined ourselves as rational creatures (surely one of the least modest statements in intellectual history)", he has written, reminding us that Freud himself saw civilisation as resulting from a suppression of the biological inclinations. "It is impossible to overlook the extent to which civilization is built up upon a renunciation of instinct", Freud wrote in *Civilization and Its Discontents*[15].

Perhaps if the answer does not lie in our capacity for reason, it lies in our basic psychology. Any psychologist will attest that humans crave certainty and purpose in life. As Edward Wilson himself admitted in *On Human Nature*: "In the midst of the chaotic and potentially disorienting experiences each person undergoes daily, religion ... gives him a driving purpose in life compatible with his self-interest". As the psychologist Nicholas Humphrey notes in *Soul Searching*: "People are bound to dream of a less threatening, less lonely and more obviously ordered world". A point that Maynard Smith and Szathmary also bring out is the enormous emotional commitment to any set of beliefs. But what originally gave us this craving for certainty and order, a craving that seems to me to be intimately linked to the ability of culture to overwrite biology (or, in Dawkins' terminology, what gave us a brain susceptible to mental parasites)? Such susceptibility to psychological tinkering may have either an adaptationist root, or may be little more than a biological

---

[15] I find it ironic that Freud, this cocaine-endorsing, slightly dodgy Viennese psychoanalyst today much maligned by figures including the evolutionary psychologists, actually had a better grasp of orthodox evolutionary theory than many in "E.P." do. In the wonderfully titled *The Future of an Illusion* Freud noted that civilisation depended upon the suppression of instinct, including the instinct for cannibalism.

A world without compassion

accident. Maynard Smith, writing with Szathmary, tentatively posits the former position, when he says that "The innate capacity to be influenced by ritual may have been individually selected, although we are not clear why it should be so". The latter position involves what is known as the "emergent properties" argument. As Williams tells us: "[a]ll through evolutionary history there have been such changes with important future consequences entirely unrelated to the selfishness that brought them about". Perhaps the great psychological appeal of certain worldviews is a necessary side consequence of having a very large brain. The only brain in the animal kingdom capable of understanding and ruminating upon its own forthcoming extinction and its individual insignificance in the great cosmic process.

Manipulation

Yet there is also another biological attribute that may hold an answer for us. Not all altruism in the non-human world can be put down to kin selection and reciprocal altruism. There is also the biological mechanism known as "manipulation", or, more accurately, "manipulating others to behave altruistically towards you".

The most celebrated example of manipulated altruism is the cuckoo. Cuckoos lay their eggs in other birds' nests, taking advantage of the fact that certain other bird species are unable to spot the intrusion and will thereafter treat the hatched chick as one of their own. From natural selection's point of view the behaviour of the surrogate parent is, in one sense, ludicrous. Here is a mother bird expending considerable effort feeding and protecting an unrelated hatchling. The reason why it works, of course, is that the genetically coded behaviour that usually works so well in raising offspring (such as: bond with whatever hatches in your nest; invest more effort in the chicks more likely to survive) has been brilliantly taken advantage of by the cuckoo. As Dawkins tells us, manipulation can evolve because animals make decisions by "rules of thumb". Everything in nature has a cost, and combating such manipulation by cuckoos will be an expensive exercise that relies on a benign mutation to get started. In this evolutionary arms race of manipulation and counter-strategy the results can be extraordinary. Given that cuckoo chicks usually get to grow in nests of physically smaller bird species, it is common to see the surrogate mother almost killing herself trying to feed the enormous cuckoo chick. An enormous chick which destroyed her natural offspring within just minutes of hatching from its own egg.

Another example given by Mark Ridley and Richard Dawkins is where predators avoid expending costly energy by manipulating their prey to cause the prey to come to them. "[T]hus are male fireflies lured by the false sexual flashes of *femmes fatales*, who eat them". Maynard Smith and Williams both tell us about another variety of manipulation – enforcement – where, for example, often the threat of physical force is sufficient to make a subordinate monkey in a group display the technical definition of "altruism" by relinquishing a feeding site to the more dominant member.

In his letter to me regarding my *Philosophy* paper, George Williams noted that manipulation should perhaps have been mentioned earlier as the third mechanism in animal altruism. I had actually thought long about this point when writing the paper. I decided not to bring it up until later in the article for two reasons. The first is that manipulation is a mechanism that gets much less press than the other two. This is largely because it is only relatively recently that it has started to be seriously studied and behaviours so categorised. As Williams notes, "Richard Dawkins [in his 1982 *The Extended Phenotype*] argues convincingly for the prevalence of manipulation in nature".

There was a second reason, though. I knew that I could probably get away with not bringing up manipulation until later in my paper because of a rather curious oversight by most sociobiological authors. Despite producing whole chapters explaining both kin selection and reciprocal altruism most spend little or no time discussing biological manipulation, and so I knew that none of these writers could upbraid me on this point. Indeed, many sociobiologists seem positively spooked by the whole mechanism. As Helena Cronin wrote in *The Ant and the Peacock*: "Finally, a more sinister explanation of self-sacrificial [*animal*] behaviour. We must consider the possibility that the behaviour really is self-sacrificial, that of a victim" to manipulation. Despite spending three short pages discussing manipulation in the animal kingdom, it is odd that manipulation does not even get a look-in when we come to the section on the biological basis of human altruism. And yet this is perhaps not so surprising; the distinguishing feature of all sociobiologists and evolutionary psychologists is their driving need to show humans as the product of their genes. But if human altruism is some derivative of

A world without compassion

biological manipulation? Well, this destroys their entire agenda in one fell swoop. Sociobiology began with the overt, stated, purpose of denying the malleability of the human animal, of putting behaviour down to genetic adaptation, and of trying to make the social sciences another part of zoology. Yet if we are being manipulated into displaying anti-adaptive behaviours, if the key is that we are being, as Williams says, manipulated into "overcoming billions of years of selection for selfishness", then we have demonstrated the malleability of the human animal. Indeed, sociobiologists would then have done more than anyone in the last hundred years towards establishing that separate sphere of the social sciences. Suddenly the idea that our behaviours are coded into our genes is gone if we act as we do simply because we dance, not to our own genes' tune, but to another's tune. Evolutionary psychologists tend to ignore manipulation because their own worldview offers them no choice but to ignore it.

I have, though, found one pop sociobiological text that - while holding to the party line that humans evolved morality as a biological adaptation - does admit a potential substantial role for manipulation in human affairs. It was written in 1991 by a lecturer in sociology at the L.S.E. called Christopher Badcock. Most of what is in the book has been put forward numerous times by better writers like Gribbin and Matt Ridley, but his work did at least posit something new for a sociobiologist. After giving us the obligatory hundred page spiel about kin selection and game theory models of reciprocal altruism, Badcock then started talking about "induced altruism", or the third force of manipulation. This, he suggested, might be "true" human altruism (although he was trying to base it all on deceit or enforcement – cuckoos or monkeys – in other words normal biological manipulation, rather than the psychological manipulation of Dawkins' "memes").

Darwinism tells us that humans are undoubtedly being manipulated. We do not need to fear the idea (as Cronin appears to), because had something *not* manipulated us free from our genes we would still be living in groups of one hundred and grunting. Yet as to whether biological manipulation has much to teach us about human manipulation, that I do not know. I tend to think probably not. Psychological manipulation is so very different from the physical or unconscious manipulation of nature. Yet I am prepared to be converted, largely because of the standing of those who hazard a connection.

George Williams has drawn a tentative link, and even Richard Dawkins sees value in starting the debate. As he wrote with Mark Ridley in their 1981 paper: "Of all the subjects in this essay, it may be manipulation that holds the best possibilities for useful interaction between evolutionary ethology and social psychology".

The resolution to Huxley's Paradox may lie in manipulation, or it may lie in one or more of the suggestions in the preceding section. Darwin himself saw a combination of factors as crucial. But we know that there must be an answer, because we have a very large gap to bridge once we cut the gene down to size. Once we have the courage to accept that we are, in fact, genetically, apes.

"this gradual strengthening of the social bond"

A point not to be missed, of course, is that as soon as culture *has* the power to act on the human mind and create group sizes never before witnessed in diploidy nature, then almost inevitably culture *will* so act. Large-scale civilisation will automatically begin to develop at some point (a point which has implications for chapter 10). Darwinian logic takes over, as selection between cultures comes into play and the more cohesive belief systems prevail. Darwin and Huxley both noted the effect, while Wallace, too, recognised the process of selection between cultures once something has first managed to overcome the small-scale selfishness of diploid nature.

In *The Descent of Man* Darwin commented that "[s]elfish and contentious people will not cohere, and without coherence nothing can be effected. A tribe possessing the above qualities [*of sympathy, fidelity and courage*] in a high degree would spread and be victorious over other tribes". Wallace in "The Origin of Human Races" had previously written: "Tribes in which such mental and moral qualities were predominant, would therefore have an advantage in the struggle for existence over other tribes in which they were less developed, would live and maintain their numbers, while the others would decrease and finally succumb". And, in 1894 in his "Evolution and Ethics" lecture, Huxley was to note that: "Other things being alike, the tribe of savages in which order was best maintained; in which there was most security within the tribe and the most loyal mutual support outside it, would be the survivors. I have termed this gradual strengthening of the social bond, which, though it arrests the struggle for existence inside society, up to a certain point improves the chances of society, as a corporate whole, in the cosmic struggle [*of natural selection*] - the ethical process".

"The Huxley-Williams nature-as-enemy idea"
...the first man to construct a state deserves credit for conferring very great benefits. For as man is the best of all animals when he has reached his full development, so he is worst of all when divorced from law and justice.

Aristotle – The Politics

Humans are not born, they are made. In his 1998 letter to me, attached in the appendix, George Williams describes it as "the Huxley-Williams nature-as-enemy idea". It is an idea also shared by Darwin, Wallace, Maynard Smith and Dawkins. Yet it is more than just an "idea". It is the only conclusion that evolutionary logic will allow us. Man is not born because "man" does not even exist until culture has defeated the "enemy within". Our own DNA is the enemy of morality, the enemy of civilisation, the enemy of everything we strive towards and have built over the last ten thousand years. Nature is the human enemy, but nature is within the nucleus of every cell in our bodies. Nature is the enemy, and culture is the saviour; be it a precarious and terrible saviour. In its full, Darwinian, understanding this idea is truly shocking. Yet the tremendous power of culture to mould individuals is knowledge as old as man himself. Humanity has a very long tradition of trying to use, or trying to study, this reprogramming. It is at least as old as the Biblical idea that we are born "fallen", but that religious reprogramming, reinforced by ritual, can set us "free". And is this ancient idea really so very different from saying that we are born apes, but that cultural manipulation lets us rise above our Darwinian past? This concept of wanting to, or needing to, reprogramme is a concept found in different arrangements in almost all thinking, Western and Eastern, secular and religious. For example, Confucianism with its emphasis on habit, ritual and peer pressure: "Guide them by edicts, keep them in line with punishments, and the common people will stay out of trouble but will have no sense of shame. Guide them by virtue, keep them in line with the rites, and they will, besides having a sense of shame, reform themselves" (*The Analects*). Or Plato's *Republic* with his famous Foundation Myth (or "Noble Lie"); of a society structured into relatively content castes through the inculcation, and constant repetition, of a powerful mythology. Or modern states, which make their children pledge allegiance to a flag each day, using ritual and repetition, and pageants with special costumes and music, as the tools of inculcation.

The individual is prepared by the sacred rituals for supreme effort and self-sacrifice. Overwhelmed by shibboleths, special costumes, and the sacred dancing and music ... he has a "religious experience". He is ready to ... die for God and country. ... Human beings are absurdly easy to indoctrinate – they seek it.

E. O. Wilson – Sociobiology: The New Synthesis

Aristotle was both very right and very wrong. Man in a "good" society can be without doubt the best of all animals, capable of incredible kindness and altruism. Also (and as we shall consider in detail in chapter 9), man in a "bad" society can be far and away the worst of all animals, capable of savage torture, rape and infanticide that temporarily can far exceed the horrors of the natural world. But man divorced from society is not the worst of all animals. Man divorced from society is just another animal. Just another selfish ape, no better, and no worse, than any other.

Man divorced from his myths, his rituals, his cultural manipulation, his education, is a selfish ape. But with these things man can become the moral being who has risen above his Darwinian past, and who is capable of so much more, both good and bad.

5

Casualties of war

Maynard Smith, Williams, Hamilton, and Dawkins ... have largely eschewed the deeply unpleasant task of pointing out more egregious sins in the work of those who enthusiastically misuse their own good work.

Daniel Dennett – Darwin's Dangerous Idea

But why has no one been willing, or able, to draw attention to the immense difference between the answers of the evolutionary psychologists, and the answer of the leading gene-selectionists? And why have the leading gene-selectionists largely eschewed the deeply unpleasant task of pointing out the more egregious sins in the work of these heretics? "The first casualty when war comes is truth", said the US senator Hiram Johnson in 1917. And there has been a war going on in evolutionary biology for over a quarter of a century.

Gene selection versus "hierarchy"
This is a war over a number of issues, including whether evolution works

by smooth gradual change or by discontinuity and sudden starts, and the relative importance of adaptation over nonadaptational biology or evolutionary by-products. However, most fundamentally the war remains over the levels at which natural selection can act - is it (almost) all about genes, or is it not?

We now understand the basic gene-selectionist position. Selfish genery sees the gene as the fundamental unit of selection and tries to explain evolution in terms of changes at this lowest possible level. The other side in this war sees things rather differently, noting that Darwinism has done very nicely for one hundred years basing itself at the level of the individual rather than its component molecules. In *Ever Since Darwin* the late Stephen Jay Gould puts his criticism as follows: "With supreme confidence in universal adaptation, [gene-selectionists] are advocating the ultimate atomism - reduction to a level even below the apparently indivisible individual of Darwin's formulation. … Here genes themselves are generals in the battle for survival". Furthermore, if one accepts that genes must co-operate to efficiently "run" an individual, why cannot one take co-operation to even higher levels - to a hierarchy of levels? Hierarchical (also known as "multilevel selection") theorists argue that group selection, properly understood, is an important part of the evolutionary process. The three most prominent theorists from within this alternative tradition are Gould, the geneticist Richard Lewontin and the paleontologist Niles Eldredge. In this debate Lewontin has tended to concentrate on combating universal adaptation, while Eldredge was the co-originator with Gould of the 1972 anti-gradualist theory of "punctuated equilibrium".

It is important to point out that it isn't right to say that gene-selectionists argue that group selection cannot occur. What gene-selectionists argue is that group selection is likely to be a very weak force in evolution because the conditions required are so onerous. A group consisting almost entirely of altruists *can* do better than a group consisting entirely of selfish individuals, because the altruism benefits the entire group. The selfish group won't have this advantage. But, as George Williams and John Maynard Smith independently demonstrated in the mid 1960s, the problem is explaining how a group consisting entirely of altruists could have come about in the first place. In a group consisting of both altruists and selfish individuals the altruists will be eliminated by within-group selection because of the problem of subversion from within. It is theoretically possible that a wholly altruistic group could arise if the group was small in number. In such a case, random genetic drift could hypothetically establish altruism within such a population. However, it could only be immediately maintained if there is little migration in from normal selfish groups, otherwise the new immigrants would end up driving the altruists to extinction.

For group selection to be a serious force in evolution it would require between-group selection - where the altruistic inclination can be favoured because it would favour the group in competition with other groups - to be stronger than any within-group selection against altruists. In the mid 1970s Bill Hamilton took the work of Williams and Maynard Smith further to show that the force for group selection would only be able to continue indefinitely if there was periodic re-assortment of the various groups such that altruists could be re-concentrated in some groups. If this didn't happen such altruists would inevitably be eliminated by within-group selection. The problem is coming up with a process of re-assortment that allows altruists to recombine in such a way that they will not immediately be taken advantage of by non-altruists.

There are two reasons why I will not go into any further detail on the genic selection/hierarchy debate. Firstly, it is far better detailed in just about any evolution book that you can pick up off the shelves of a *Waterstone's* than I could ever manage. Writers from either tradition can give you a very good grounding in the basic points of contention (though each tradition's adherents naturally tend to explain their position a little better than they may take note of the other tradition's concerns). Secondly, I have made the mistake of being drawn into this endless debate once before, when the debate actually had nothing to do with the central argument within a paper I had submitted.

So this should be understood as a book written by a dyed-in-the-wool gene-selectionist, and it takes as its starting point the (here necessary) assumption that gene-selectionism is substantially correct. This does not mean that I see no value in the work of many of the multilevel selection theorists; on the contrary I have great respect for a number of them. They produce important work in their own right, and they have also been very important in reining in some of the more extreme formulations of gene-selectionism. Their criticisms have helped make gene-selectionism what it is today, and by continuing to point out areas of contention that gene-selectionists might like to brush over they will no doubt continue to help us all reach a better understanding of evolutionary forces. I applaud the measured tone of George Williams who has given credit where credit is due, and has sought to build bridges with the hierarchicals. (Williams has also pointed out, as others have done, that many of the disagreements between the two traditions are ultimately relatively minor in substance). I dislike the personal attacks that are made on hierarchicals by certain gene-selectionists; equally I have little time for the personal attacks made on gene-selectionists by hierarchicals[16].

This book has only one goal: to make people understand that contemporary selfish gene-ery shelters two quite separate factions, one orthodox and one heterodox - those who wish to treat us as apes, and those who do not. And to accept this you do not have to take any view whatsoever on hierarchical theory (good or bad); you only have to understand that the gene-selectionism / hierarchy war has helped keep the immense differences between the two gene selection factions hidden from view. I mentioned above that I had made the mistake of being drawn into the gene-selection versus hierarchy debate before. Over a year before I wrote "Unnatural Selection", I had made an initial stab at starting some discussion over the tremendous difference between the view of human evolution as offered by the leading gene-selectionist Darwinians and the view offered by the evolutionary psychologists. That paper was a response to the hierarchical theorist Paul Griffiths' paper in the March 1995 issue of *The British Journal for the Philosophy of Science*. Griffiths' paper, "The Cronin Controversy", used a review of Helena Cronin's *The Ant and the Peacock* to discuss various issues raised by the book. "The Cronin controversy was, and is, a dispute between two current schools in biology and the philosophy of biology", read his introduction. My Reply noted that when it comes to the understanding of Darwinian principles themselves this is a perfectly adequate characterisation. However, when it came to understanding the human animal it would not do. "In applying these initial Darwinian principles to mankind we see some fundamentally different approaches. The purpose of this Reply is to raise awareness that the gene selection school must not be thought of as an homogenous set when this

---

[16] The journalist Andrew Brown's 1999 book *The Darwin Wars* is a good review of the personalities and the unhelpful comments that have been made by both sides.

application occurs." Paul Griffiths was asked to referee my article by the editors of the *BJPS*, and he was more than fair to me. "The criticisms of my review article on Cronin are generally well-taken, and they raise interesting issues," he wrote. He would have been strongly inclined to recommend my paper's publication, Griffiths told the *BJPS*, had I not unfortunately made serious errors in my characterisation of his school. I was asked by the editors to respond to Griffiths' objection, and I amended my paper to take account of his comments, but left in a rather unwise remark. I suggested that hierarchical theorists may have less of an issue with gene-selectionism once they understand the implications of selfish gene-ery for the human animal stripped clean of the unorthodoxies of writers like Wilson and Cronin. This was a mistake. Hierarchicals, rather understandably, don't like to be told that they themselves might abandon most of their objections to a competing worldview once we simply agree on the human animal. I later tried to back-peddle, removing all references to hierarchical theory, and pointing out that my paper was solely about heterodoxies within selfish gene-ery, but by then it was too late. The damage was done, and the paper was effectively dead.

The point of the above was to explain where I come from, and where I am going (for those still in doubt). My targets are restricted to the heretics who claim to be within selfish gene-ery, and the point of this chapter is to provide some explanation as to why the leading gene-selectionists have so easily accepted the heretics as, if not brothers, at least cheeky neophytes.

My enemy's enemy ...
My deviation about Griffiths' article now helps serve a second point. I briefly scooted over the title of Griffiths' paper, "The Cronin Controversy". Remember Dennett's quote that introduces this chapter? "Maynard Smith, Williams, Hamilton, and Dawkins ... have largely eschewed the deeply unpleasant task of pointing out more egregious sins in the work of those who enthusiastically misuse their own good work." Why should such a scientific duty be unpleasant? Because professional evolutionary biology is a very small arena, and one must always learn to identify the real enemy.

When Helena Cronin was sent for a period by the LSE to Oxford University to help her gain background for her Ph.D., she began to move in the tight circle of academic selfish gene-ery. She (like the science journalist Matt Ridley) became a student of Dawkins' at Oxford. The book that resulted from her Ph.D., *The Ant and the Peacock*, had a foreword by John Maynard Smith. She is gratefully thanked by Dawkins for her help in updating *The Selfish Gene*. George Williams describes her as a "good friend" in *Plan and Purpose in Nature*. She is, above all, a hard-working ally in the war against the hierarchicals. As Paul Griffiths noted for us in an earlier chapter, group selection is the "villain" of much of *The Ant and the Peacock*. She is a loyal foot soldier helping to carry the banner of gene-selectionism to the wider world. And if some of her thoughts are a little ... *unorthodox*, well, surely they pose less of a threat to orthodox Darwinism than those dreadful hierarchicals? You see, although the leading gene-selectionists fear the damage that the evolutionary psychologists are doing to Darwinism, they fear losing ground to the hierarchical tradition even more. And so selfish gene-ery never apologised for the errors of the earlier human sociobiology (even though it was believed to speak in its name), and neither did they expel its advocates from the school. This would have given much needed ammunition to the hierarchicals. After spending decades establishing themselves as the premier school, the gene-selectionists could not afford such a public humiliation. In war you do not help your enemy. And in this war human self-knowledge was an acceptable casualty. Sociologist Ullica Segerstrale claims that Ernst Mayr, one of the founders of the "Modern Synthesis" of Darwin and the genetics of Mendel in the mid twentieth century, admitted to her that it would have been "so easy" to criticise sociobiology on purely scientific grounds. Mayr knew of at least three leading biologists who had initially been severely critical of human sociobiology but "tore up" their criticisms planned for the peer-reviewed scholarly journals because they were nervous of being seen as in any way supportive of the hard line views of Gould and Lewontin.

This perceived need to constantly attack, or at least never support, those within the hierarchical tradition can reverberate throughout other disciplines. The palaeontologist Simon Conway Morris produced an influential book a few years ago analysing and interpreting the fossils from the famous Burgess Shale in British Columbia. So it was fascinating to come across Richard Fortey's take on things. Fortey is a senior palaeontologist at the Natural History Museum in London, and his wonderful book on trilobites was a real eye-opener in many ways. In *The*

Casualties of war

Crucible of Creation Conway Morris had savaged Stephen Jay Gould's account of the Burgess Shale investigations and the supposed Cambrian "explosion" of life. "Savaged" is an apt term; Fortey writes that he had "never encountered such spleen in a book by a professional". Fortey's rationale for the spite is worth reading. Suffice to say, according to Fortey (and I have no reason to doubt his account), the venom was undeserved by Gould, who had been nothing but honourable in his treatment of Conway Morris. But what really grated with me was Dawkins' apparent lauding of Conway Morris *simply because he was attacking Gould*. Fortey certainly appears to believe that some of those, like Richard Dawkins, who drew attention to Conway Morris's criticisms of Gould, may not have been fully grounded in the history of the "explosive" opinions (2000, pp. 135-8). Opponents of Gould in other arenas, Fortey wrote, they appear to have used the book as a stick to beat him, "operating on the principle: 'my enemy's enemy is my friend'".

For the evolutionary psychologists themselves, both the friendships and, more importantly, the ally status reap rewards. In 1992 and 1993 "The Cronin Controversy" erupted in the pages of the *New York Review of Books* (where so much of the recent Darwinian vitriol has tended to appear). Gould slammed Cronin for both being a gene-selectionist and for jealously co-opting human explanations into orthodox gene-selection. Yet Cronin's teachers could see only another attack on selfish gene-ery. Maynard Smith and Dennett both leapt to the defence of gene-selectionism, while the fundamental question of what Darwinism can tell us about the human animal - for Darwin "the highest & most interesting problem for the naturalist" - was almost ignored. Maynard Smith did admit he had important reservations over Cronin's human interpretations: "As it happens, I agree with [*Gould*] that there is more to the evolution of human altruism than kin selection: once a species has acquired language as a second method of passing information between generations, new mechanisms of change become possible." Nevertheless, this important point was again lost in the same tired ding-dong altercation between gene-selection and hierarchy where circling the wagons, and not furthering human self-understanding, appears to have become the prime objective[17].

---

[17] According to Segerstrale (2000, p. 192), long-time chronicler of the evolution wars,

Please do not ask for credit as a refusal often offends

What is particularly maddening in the endless gene selection / hierarchy debates is that all too often neither side even seems to try to understand what the other is saying. So this is Richard Lewontin in 1994 in the *New York Review of Books*: "[t]here is, at present, no aspect of social or individual life that is not claimed for the genes. Richard Dawkins' claim that the genes 'make us, body and mind' seemed the hyperbolic excess of vulgar understanding in 1976, but it is now the unexamined consensus of intellectual consciousness propagated by journalists and scientists alike". Now Dawkins is absolutely right, genes do (in a sense) make us, body and mind. A newborn infant is the product of nothing more: as Lewontin does accept, it cannot be the product of anything else since only genes (and a few maternal hormones) pass physically between parent and embryo. The second inheritance system of cultural influence has not yet had a chance to work its wonders. But to so misunderstand Dawkins' *The Selfish Gene* as a hymn to the gene (to ignore, in effect, the final chapter of his first edition) is extraordinary. Yet this is the story of the last twenty-five years, as gene-selectionists misunderstand hierarchicals, and hierarchicals misunderstand gene-selectionists. No one is given credit, and all are tarred with the same brush.

Yet the shift from human sociobiology to evolutionary psychology (a shift from ultimately saying that human differences are largely genetic to saying that human differences are almost entirely cultural) is considerable. And credit for causing this shift is very largely with the hierarchicals; precious little help was forthcoming from the leading gene-selectionists. The journalist Andrew Brown put it quite nicely in his recent book *The Darwin Wars*. "The development of evolutionary psychology from [*human*] sociobiology can be understood in two ways. Looked at purely ideologically, it is a triumph for Gould and Lewontin, who have seen almost all of their original objections incorporated into the project". But it came in a context of unremitting personal hostility to Gould, Brown notes. "Politically, however, the story seems to be one of

---

when critic of gene-selectionism Steven Rose threatened to sue Dawkins for libel in 1985, Hamilton reportedly approached E.O. Wilson in order to get Segerstrale's Ph.D. thesis ("Whose truth shall prevail? Moral and scientific interests in the sociobiology controversy") for use in a possible legal defence of Dawkins against Rose.

the steady marginalisation of [*the hierarchical school*] and the emergence and triumph of a refined and purified sociobiology." Gould himself is particularly loathed[18], Brown explains, even though some of his criticisms of Wilson's original *Sociobiology* are now accepted even by its sympathisers. Brown interviewed Cronin, who reportedly commented that parts of *Sociobiology* "were not even testable", while others were "simply false, and, interestingly, bad Darwinian theory".

The savage parochialism of all this makes me angry; yes, Gould had an often abrasive character and difficulty admitting to his own errors and over-enthusiasm. And in that he was very like many leading gene-selectionists, especially those based in Oxford. But his contribution has been extremely important, both in his own work, and in his role of policing the work of the poorer putative "gene-selectionists". By default selfish gene-ery needed Gould for so long as its own thinkers "largely eschewed the deeply unpleasant task of pointing out more egregious sins" in the work of their less orthodox disciples. So in the end it is not hierarchy or gene-selection that truly suffer from this unnecessary hostility; it is Darwinism. There are many things that can be calmly, yet passionately, debated. But allowing heretics to set the agenda according to which we must debate "the highest & most interesting problem for the naturalist" is a shocking betrayal of Darwin.

Philosophy and the social sciences

My initial interest in Darwinian theory was roused by philosophers' criticisms – not because I thought that they were right but because I was convinced that they must be seriously wrong.

Helena Cronin – The Ant and the Peacock

It is not only Darwinism that suffers from the relentless hostilities; philosophy, too, is the poorer for it. The confusion and missed opportunities have caused friction beyond the borders of evolutionary biology, and again, there has been fault on both sides.

Dennett comments in *Darwin's Dangerous Idea* that it is the hostility of their detractors that has created a "siege mentality" within gene-

---

[18] As an example of what Gould was up against in the years before his death in May 2002 see Robert Wright's December 1999 piece "The Accidental Creationist: Why Stephen Jay Gould is Bad for Evolution".

selectionism that has led the leading theorists to eschew their "duty" to expose the sins of their less orthodox disciples. This is particularly true of the criticism that has come from outside biology. The leading gene-selectionist biologists have often run into so much hostility from philosophy and the social sciences that they have usually found it easier to avoid discussion altogether[19]. A good example of this was the animosity of some philosophers and social scientists to Dawkins' *The Selfish Gene*. In 1979 the philosopher Mary Midgley wrote a paper published in the journal *Philosophy* criticising, amongst other things, Dawkins' use of metaphor (a device he often uses to get a larger message across). The paper was quite extraordinarily unfriendly; *Philosophy* permitted Dawkins a reply, published in 1981, and his bitterness was manifest: "such transparent spite .... so rude .... hard for me not to regard the gloves as off". Time has not healed these wounds, and according to Andrew Brown, Dawkins is still so offended by her attack that in 1992 he withdrew from a residential conference once he heard that Midgley had also been invited. Now I usually have a lot of time for Midgley. At times she seems to be fighting a lonely battle to rid philosophy of its focus on self-important logic chopping and is trying to make it more relevant to the outside world. She's a very good philosopher, but she is not a philosopher of science. And such intemperate attacks on major biologists have not helped things.

Perhaps worse than intemperate attacks, however, are philosophical tracts that support a naturalistic view of ethical man while seeming to make little attempt to understand the subject they are commenting upon. In *Philosophy* I highlighted the political theorist Allan Gibbard's inability to keep distinct material differences of opinion between the different evolutionary traditions. Gibbard's *Wise Choices, Apt Feelings* was his attempt to provide a naturalistic theory of normative judgement. In contrast Michael Bradie's *The Secret Chain* was a wide-ranging

---

[19] The hostility exists on both sides, of course. The influential geneticist Steve Jones made the following unhelpful comment in the *NYRB* in 1997: "For most wearers of white coats, philosophy is to science as pornography is to sex: it is cheaper, easier, and some people seem, bafflingly, to prefer it". By the end of this book I hope you will understand why both human morality and self-knowledge depend on scientists, philosophers *and* (after chapter 8) theologians working together, simply because by working independently each group has been found sorely wanting.

investigation of how both altruism and ethics have been rationalised from the seventeenth century onwards, and a detailed assessment of several contemporary sociobiological approaches. Bradie concluded that evolutionary considerations alone are not capable of providing a foundation for ethical theory, but stressed that a Darwinian perspective is capable of generating potentially deep insights into the nature of our moral practices. Following fellow American John Dewey, Bradie welcomed the important perspectives evolutionary theory may bring to old questions and the light it may shed on moral justifications.

Bradie knows he must make an accommodation with evolutionary theory, but he, too, took human sociobiology to be largely synonymous with selfish gene theory. "Most contemporary sociobiologists, if not all, are so-called genic selectionists", he wrote. Yet the biologists who actually developed the theory of genic selection back in the 1960s and 1970s are given short shrift. Despite repeated calls from Bradie that we should listen to the evolutionary biologists, Williams' support for Huxley's stance is reduced to one sentence of his book, Maynard Smith is not even consulted, and the author of *The Selfish Gene* is placed on the "Hobbes, Mandeville, Dawkins ... side". The objects of human desires for the seventeenth-century philosopher Thomas Hobbes were self-regarding pleasures, Bradie explains. All apparently benevolent and altruistic impulses have to be understood in terms of disguised self-interest. He then told us to "see, e.g., Dawkins (1976) ... for a contemporary sociobiological variation on this theme". Dawkins' crucial point that the *third* major route to altruism in nature, manipulation, is the key to understanding much human altruism, was entirely overlooked by Bradie. Any value in Bradie's work was fatally undermined by his inability to get to grips with the work of the leading selfish gene theorists, a weakness that marks philosophical discussion of this subject time and again.

However, other philosophers have tended to approach human biology with a different agenda. Two major philosophers to have produced serious thought on the ideas being put forward by the human sociobiologists have been Roger Trigg and Peter Singer. Both accepted that ultimately we are the products of natural selection. Both therefore intelligently recognised early on that the social sciences had to come to some sort of accommodation with human evolutionary biology, and could not simply dismiss it out of hand as social scientists had previously sought to do. If human sociobiology was largely wrong, reasons had to be given for why it was wrong.

Trigg, in his 1982 *The Shaping of Man*, generally seemed to be claiming that the problem was that sociobiology was too ambitious in believing that it could explain the complexities of the human mind. This is potentially a hazardous route to go down (many mammals have very complex brains), and his later thinking gets closer to the nub of what is wrong with human sociobiology. In a short chapter in a 1988 work he wrote: "Once, however, a tribe grows into a nation, which is not composed of near relatives ... it is questionable how far a highly developed 'moral sense' could be under genetic control". Yet by never actually taking Darwinism back to its roots and understanding it the way that Dawkins, Maynard Smith and Williams understand it, Trigg has never been able to make the final connections. The bigger question for Darwinism is how the tribe can grow into the nation in the first place. Nevertheless, as an attempt to at least begin to fight on the evolutionary battlefield it was laudable in its intentions. Of course, the social sciences' more commonplace attitude of ignoring sociobiology is, in one sense, understandable, because the sociobiologists often never had substantial or meaningful empirical data on their side. The sociobiologists have never really bothered to attempt to engage on the social scientists' own field of battle, so the argument has been made as to why social scientists should have to be the ones to engage on both fields of conflict. But it is also an intellectually vacuous position. You do not win wars by refusing to fight the enemy on his home ground when it is his home ground that holds the key advantages (popular appeal amongst them). If social scientists are convinced of their positions then they should (like Roger Trigg) be looking to the science itself to understand why human sociobiology has been going so very badly wrong. Now this is of course easier said than done in the modern world of academic specialisation, but nevertheless the answer has been out there for over thirty years in the work of the leading gene-selectionists.

The Australian philosopher Peter Singer tried to be more accommodating than Trigg to sociobiology, but ultimately concluded that reason set us free from genetic control - that reason masters our genes (an important argument that we shall consider in a later chapter). We go beyond the altruism of the natural world because we have "reasoned" altruism, says Singer. But the danger of trying to accommodate human sociobiology without first fully mastering the underlying science is shown by the loose way he expands from this "type" of altruism to "true" altruism. Members of the European Sociobiological Society read the following in a review of his book: "Fortunately, the author has succeeded in his assimilation of [*human*] sociobiology. ... Yet, to your reviewer he leaves some questions open ... [*and*] he puts sociobiological theory in doubt because 'genuine, non-reciprocal altruism directed toward strangers does occur'" (Wind [1984]).

This was still a commendable effort, though, since most academics outside biology have not even seen the need to attempt a cautious (albeit somewhat flawed, in the case of Singer) analysis of human Darwinism. Most social scientists, and many philosophers, see the enormous gap between the pronouncements of a Wilson, or a Pinker, or a Cronin, and the long-held beliefs within orthodox anthropology and psychology, and simply turn away. This is understandable, but this is also unacceptable.

You want to live "according to nature"? O you noble Stoics, what fraudulent words! Think of a being such as nature is, ... think of indifference itself as a power - how could you live according to such indifference? To live - is that not precisely wanting to be other than this nature?

Friedrich Nietzsche – Beyond Good and Evil

"indifference itself as a power"

Selfish gene-ery has been the orthodox interpretation of Darwinian theory for well over two decades now, and in all that time it has never faced a serious threat to its hegemony. Yes, it has had to make accommodations, and yes, it has had to draw back on over-enthusiastic claims, but this is simply the normal process of scientific advancement. It was never really facing any fundamental challenge from religion, philosophy, or hierarchical theory, as those within the tradition have always known.

The great tragedy is that for twenty-five years selfish gene-ery has proceeded as if it were under mortal attack, and has consequently gained a harmful fundamentalist image, and a tendency to shield the heterodox at the expense of those outside the tradition altogether. The tragedy has been that by eschewing their scientific "duty", as Dennett

Casualties of war

calls it, leading Darwinians have sacrificed human self-knowledge to win a war that they had already won. Yet the cost of the unnecessary hostility has been the non-acceptance of modern Darwinism by philosophy and mainstream social science. The cost has been to defer all proper investigation into "the highest & most interesting problem" in biology. So at the end of the day the real loser has been Darwin.

6

Blessed of Nature

It is important to recognize that Darwinism has always had an unfortunate power to attract the most unwelcome enthusiasts. ... Gould has laid this sad story bare in dozens of tales, about the Social Darwinists, about unspeakable racists, and most poignantly about basically good people who got confused - seduced and abandoned, you might say - by one Darwinian siren or another.

Daniel Dennett – Darwin's Dangerous Idea

Many have pointed to the openly political statements of some big name sociobiologists. In *On Human Nature* Wilson wrote that "Marxism ... is now mortally threatened by the discoveries of human sociobiology". Wilson has always held that his science had led him to his politics, and not the other way around. That he had no option but to see the light. "[A]nd I have always said – although some of my critics didn't listen – that had the evidence emerged that a socialist society was the one most congenial to the basic property of human nature, then I would have voted Socialist" (in interview with Radford [1995]). Wilson is arguing that while left wing doctrines are undermined, the Right is supported by Darwinism because both *laissez-faire* economic self-interest and group allegiance or "duty" are products of nature. Such behaviour *is* our selfish genes manifesting themselves. In other words, Neanderthals would have grunted *The Star-Spangled Banner* if they could only have carried a tune, because both patriotism and capitalism are natural to our ape line. Today evolutionary psychologists like Matt Ridley and Steven Pinker come to similar scientific conclusions.

The idea that scientists or social scientists operate in a vacuum immune from their prior political beliefs is absurd. The *best* scientists will always struggle hard to keep their personal beliefs and their professional studies separate, but few will succeed to any great extent. Not only will their views colour their choices of study in the first place, and their interpretation of data, but the idea that most will even *want* to try to separate the two is naïve. And we should be beginning to understand why this is the case. Not only are our core beliefs our way of anchoring ourselves in a frightening world (giving each "a driving purpose in life compatible with his self-interest", as Wilson told us), but they will have been inculcated so deeply that most of us cannot even begin to accept that our core beliefs may actually be wrong in major respects (we shall discuss the problem of "self" in a later chapter). This does of course bring with it both methodological and epistemological problems. In many cases those we *should* most trust to speak will be those whose sphere of study has *few* implications for the human animal. Even then their work may not be untainted, but at least we should walk in less fear of ideological involvement. The philosopher of science Karl Popper realised this, but his methodological understanding (which is still today seen as the paradigm of good scientific method) hopes to largely get around this danger. In his 1961 *27 Theses* he says that it is a serious mistake "to assume that the objectivity of a science depends on the objectivity of the scientist. ... The natural scientist is just as partisan as other people. ... [*Objectivity is*] rather the social result of their mutual criticism, of the friendly-hostile division of labour among the scientists, of their competition" (taken from Douglas Williams' *Truth, Hope and Power: The Thought of Karl Popper*). There are of course many people who dispute that this ideal is often anything more than an ideal, but if we assume that for the large part it holds true, then we can see that the most dangerous time for science is when it decides to look the other way. Such as when those with the greatest knowledge eschew the deeply unpleasant task of scientific criticism.

A short history of eugenics
Eugenics is the theory that the human race can be improved by controlled breeding from chosen individuals, with the objective of selection for, or elimination of, specific characteristics. Probably the first person to carefully articulate the idea was Plato two and a half thousand years ago with his ideal society as set out in *The Republic*. In Plato's utopia, within his governing classes chosen men and women would breed under controlled conditions (or "mating festivals"), and their children would be brought up in state nurseries. Plato saw a considerable role for culture in the human make-up, so his system of eugenics was designed as much as a way to smash the family (to prevent loyalty to the state being harmed by interpersonal loyalty) as it was to breed better individuals. Plato's motives were complex and are still today hotly debated (in part he was one of the first people in history to desire sexual equality, as he wanted women to be as "free" as men because he saw them as capable of occupying exactly the same positions). Nevertheless, Plato still saw eugenics as a way to improve the health, strength and intelligence of human stock. Indeed, a level of eugenics was actually practised by certain Greek states at this time, and at least some of Plato's ideas were drawn from the controls used in the military states like Sparta.

The term "eugenics" itself was coined in 1883 by Francis Galton, a cousin of Charles Darwin. Early in his writing career Galton stressed the biological inheritance of every kind of moral and mental trait, from low to high intelligence, to aggression, criminality and drunkenness. Galton believed that not just individuals but groups and nations inherited their traits. Perhaps unsurprisingly, for Galton the white, Victorian, male elite was nature's highest human achievement. And since Galton eugenics has found favour with notables from the leftist George Bernard Shaw (who, like many of his period, supported the active extermination of those who did not possess the desirable qualities) to Neville Chamberlain and even Winston Churchill. Today there are still those who know who should be allowed to breed, and who should not be. In 1995 the Chinese government passed a eugenics law that allows the state not only to genetically screen couples who wish to marry, but also to determine what are to be considered as "inappropriate" genes.

The hierarchical theorist Richard Lewontin has done as much as anyone to try to draw to our attention the political or ideological links of many behavioural geneticists and sociobiologists, and of the long history of politics driving biology. The first serious "scientific" study of the internal biological causes of social position was Cesare Lombroso's late nineteenth-century criminal anthropology, which claimed that criminals were born and not made, Lewontin tells us. "This theory of innate criminality, updated to implicate faulty DNA, has a modern current and, indeed, is taught at Harvard. There has been, since Lombroso, a major intellectual industry tracing the causes of social inequality between classes, races, and the sexes. A vast literature has been created" (Lewontin [1994]). Gould, too, sounds this warning from history, reminding us that Lombroso drew his conclusions directly from evolutionary biology. "[B]orn criminals are essentially apes living in our midst. Later in life, Lombroso recalled his moment of revelation: '… I found in the skull of a brigand a very long series of atavistic anomalies'." Lombroso wrote that the problem of the nature and of the origin of the criminal seemed to him now resolved. The characters of "primitive men and inferior animals" were being reproduced in the present.

Lombroso and his followers saw themselves as "criminal anthropologists"; biology could be harnessed as a legislator's tool. "Lombroso even suggested a preventive criminology - society need not wait (and suffer) for the act itself, for physical and social stigmata define the potential criminal. He can be identified (in early childhood), watched, and whisked away at the first manifestation of his irrevocable nature"[20]. Some of Lombroso's colleagues favoured extermination, rather than exile, of these natural criminals. Gould finishes by reminding us that Lombroso's work had racial implications too for its adherents: "If human savages, like born criminals, retained apish traits, then primitive tribes - 'lesser breeds without the law' - could be regarded as essentially criminal". Of course, says Gould, no one takes the claims of Lombroso

---

[20] "The Criminal as Nature's Mistake, or the Ape in Some of Us". Reprinted in *Ever Since Darwin*. Lombroso's thesis would ultimately be resurrected by the Nazis, who used it as an excuse to begin their liquidation of the German underclass. German Aryans were not natural criminals and so could be forgiven their peccadilloes; Jews and Slavs were natural criminals and so could not, went the propaganda.

seriously today: "His statistics were faulty beyond belief; only a blind faith in inevitable conclusions could have led to his fudging and finagling. ... No serious advocate of innate criminality recommends the irrevocable detention or murder of the unfortunately afflicted or even claims that a natural penchant for criminal behavior necessarily leads to criminal actions".

Gould wrote this essay in the mid 1970s, from the heady days of America's Bicentennial before evolutionary psychologists told us that there was no such thing as society. One wonders what he would have made of the geneticist John Turner's essay in *The Times Literary Supplement* in February 1999. Turner comments: "[O]ne day, we may be able to use genetic analysis to predict far in advance that someone is destined to develop a severe and dangerous personality disorder. Would such a child then be marked out for indefinite imprisonment on reaching adolescence?" It is interesting to note that while Lombroso favoured allowing these innate criminals to commit a single crime first before they are locked away forever, some modern geneticists wonder whether this important legal requirement will one day even be necessary.

Apart from the left-leaning hierarchical Darwinians like Gould and Lewontin, others have also seen fit to draw attention to both the generally right wing political views of the current crop of sociobiologists and evolutionary psychologists, and the risk of a scientist's political views colouring his or her scientific understanding. In the 1989 edition of *The Selfish Gene* Dawkins hit out at both his hierarchical critics, who are often "high priests of the left", and his other critics - those "doctrinaire sociobiologists jealously protective of the importance of genetic influence" in the human animal, and who come from the "opposite quarter" of the right. History teaches us that, for many at the political extremes (right and left), genetics is only fun when it can be used to hurt or hate others. In 1924 the United States passed the Immigration Restriction Act, severely restricting entry by the "genetically inferior" peoples - non-Europeans and those from southern and eastern Europe. In his essay "Racist Arguments and IQ" (in *Ever Since Darwin*) Stephen Jay Gould called the Immigration Restriction Act the victory of the eugenics movement, culminating as it did from years of sustained pressure by the Eugenics Record Office, run by one of America's leading geneticists, Charles B. Davenport. In preparation for the 1924 Act, many

Blessed of Nature

highly regarded biologists had sworn before Congress that Italians were (for biological reasons) more likely to be poor, and Jews were more likely to be greedy. Despite this history, in today's America the descendants of Italian and Jewish immigrants often join their fellow Americans in claiming that modern genetics shows blacks, criminals or the poor to be genetically inferior. It does not, of course, but human genetics has so rarely been about science, and so often about politics. Yet their fun will begin to pale when such people are forced to face the simple biological truth that all peoples, and all people, are born grossly immoral apes that cultures then have to re-make. Decency and kindness, like racism and discrimination, has to be learned.

A long history of hype

This theory [of natural selection] *has a history of misuse almost as long as its proper pedigree. ... Every evolutionist knows this history only too well, and we bear some measure of collective responsibility for the uncritical fascination that many of us have shown for such unjustified extensions.*

Stephen Jay Gould – "The Most Unkindest Cut of All"[21]

The history of gene research has been a history of hype. The problem with gene research is that it lends itself so easily to this kind of hype. Time after time we see exactly the same pattern: an explosive claim is made (some group or some race has this or that malign gene or specific evolutionary characteristic), it is picked up by a media interested in controversial science rather than good science, and the claim is widely discussed because it speaks to the basic prejudices of so many while drawing great hostility from the targeted group itself or its defenders. The claim thus makes its way into the public consciousness, is often accepted as another scientific "fact", but only much later falls away or is withdrawn. As Gould puts it: "Once again, biological determinism makes a splash, creates a wave of discussion and cocktail party chatter, and then dissipates for want of evidence". The problem, though, is that such claims do not dissipate for so many. The original claim will be heard around the world and is page one news for so many publications, while the retraction or scientific community's rejection after counter studies is heard by only a handful. It appears as a brief note on page 15, if it

---

[21] Re-printed in *Dinosaur in a Haystack*.

appears at all. A very good example of this trend is Dean Hamer and his "gay gene".

In 1993 Dean Hamer, an American AIDS researcher and behavioural geneticist, made a worldwide splash when he announced that his twins studies showed that gay men often carried an identical segment on their X chromosome, one of the two chromosomes that determine sex (in humans the female is XX and the male XY)[22]. He became an international celebrity, with his work celebrated equally by those who wanted to show that homosexuality was natural and thus unavoidable, and those who wanted foetuses tested for the gene and aborted if they carried it. Darwinians were found who provided evolutionary rationales for how such a gene could have been maintained within populations. Few stopped to even consider the merits of the study, or its uncorroborated status. In April 1999 a new team of scientists used a larger sample of homosexual men, and their findings (reported again in *Science*) were that the site identified by Hamer and colleagues was no more likely to be present in homosexual than non-homosexual men[23].

Now my point is not that there may or may not be a weak or a strong genetic component to homosexuality. My point is not even over whether or not the second study was better conducted than the first. In the next chapter we shall briefly consider genetic inputs to types of behaviour like homosexuality, and evolutionary theory can do little more for us than raise a few rather interesting side issues in this area. No; my point here is about the scientific process. The first study came out in a blaze of publicity, while the subsequent study was not even reported by most UK newspapers, and only *The Observer* carried an in-depth comment on the news. Broadcast media also gave it little or no airtime. And yet if the first study was indeed poor, making claims that science could not corroborate, the damage had already been done. A fraction of those who heard tell of scientists "discovering" a "gay gene" will have heard about the subsequent finding or made the link to the much trumpeted initial work.

But unfortunately this is the history of genetic claims. As the science

---

[22] First reported in *Science*, July 16 1993. Hamer, Magnuson and Pattatucci, "A Linkage Between DNA Markers on the X Chromosome and Male Sexual Orientation".

[23] *Science*, April 23 1999. Rice, Anderson, Risch and Ebers, "Male Homosexuality: Absence of Linkage to Microsatellite Markers at Xq28".

Blessed of Nature

journalist John Horgan noted in a 1993 article for *Scientific American* on behavioural genetics, the number of claims in this discipline subsequently not corroborated or withdrawn is rather amazing. Within his article, a sub-section headed "Behavioral Genetics: A Lack-of-Progress Report" notes the various findings. Twin and adoption studies for a genetic component to crime have concluded factors of anywhere between 0% and more than 50% - according to your politics or inclination you can take your pick. Widely reported research claiming a link between an extra Y chromosome and aggression was subsequently found by follow-up studies "to be spurious", Horgan notes. Intelligence (which certainly has some genetic component and which we shall consider in the next chapter) has seen twin and adoption studies showing a component to variability of anywhere between 20% and 80%. He makes similar comments for studies showing high genetic components to manic depression - "[b]oth reports have been retracted"; to schizophrenia - "the initial claim has now been retracted"; and for a gene for alcoholism - "[a] recent review of the evidence concluded it does not support a link". Horgan notes also the modern situation when so many in positions of influence seem to believe that genetics can answer all society's problems. "Daniel E. Koshland, Jr., ... editor of *Science*, the most influential peer-reviewed journal in the U.S., has declared in an editorial that the nature/nurture debate is 'basically over'. ... He has contended that genetic research may help eliminate society's most intractable problems, including drug abuse, homelessness and, yes, violent crime". I regret that so influential an editor could have publicly declared such a decided position when evolution raises so many unanswered questions. For a gene to even hypothetically exist in the first place it must be explainable under mutation and inheritance, and the nature/nurture debate is most assuredly not over unless we are to disregard the conclusions drawn from orthodox evolutionary biology.

Back in March 1998 BBC television hosted a weekend of programmes devoted to Darwin and his legacy. One of the final slots was a debate between various protagonists in the field, Steven Pinker and the noted British geneticist Steve Jones amongst them. Jones has always been sceptical of many of the claims made in human evolutionary genetics. Steven Pinker was busy enthusing about a new behavioural genetics claim just out (I can't remember which particular claim it was). The moderator (Melvyn Bragg, a keen proponent of evolutionary psychology) asked Jones to respond to Pinker. Jones launched into some complex technical argument against the claim, thereby losing half his audience immediately. Yet for the life of me I don't know why Jones' response to Pinker wasn't simply: "Oh, really? You wait eighteen months".

Misleading the debate

The extraordinary misconception of the whole subject by popular writers and reviewers is well shown by an article which appeared in the Times *newspaper. ... In this passage the theory of natural selection is so absurdly misrepresented that it would be amusing, did we not consider the misleading effect likely to be produced by this kind of teaching in so popular a journal.*

Alfred Russel Wallace – (October 1868)

The errors of human sociobiologists and evolutionary psychologists, while often based on extraordinary empirical naivety, are at least partially understandable. If we ignore the possible motivation behind such errors, human sociobiologists and evolutionary psychologists can at least claim to have been confused by the esoteric and often complex debates involved in modern Darwinism. However, the disciples of "behavioural genetics" cannot put up such a defence. Behavioural genetics, long worshipped by the political right, seems often based on a truly *awe-inspiring* rejection of common sense.

I could not believe my eyes when I read *The Times* on the 3rd November 1997. On page 15, under the headline "Twins Prove Life's A Script", we were introduced to the findings of American behavioural geneticists. In an almost word-for-word extract from the science journalist Lawrence Wright's new book *Twins: Genes, Environment and the Mystery of Human Identity* we were told the following:

Professor Thomas J. Bouchard was sitting in his office at the University of Minnesota when one of his graduate students came in with the Minneapolis Tribune. *"Did you see this fascinating story about these twins who were reared apart? You really ought to study these". Bouchard began to read the story. ... Bouchard thought it was odd enough that both were named James, but it was uncanny that each man had married and divorced a woman named Linda, then married a woman named Betty ...*

I scanned the letters page of *The Times* for days afterward, but not a single mention was made of this truly gob-smacking article. Read it again. Bouchard "thought it was odd enough that both were named James". And this fact is supposed to be of interest to a behavioural geneticist? Behavioural geneticists are interested in how a gene may manifest itself in an environment, so by displaying surprise Bouchard seems to be telling us that he believes that babies' genes somehow indicate what name they would be happy with. This appears to be the point of the comment, because the only other reading of this piece is that Bouchard (and the *Minneapolis Tribune*) was shocked that the new foster families had named both of these boys "James" in response to recognising a common "James-like" quality in the two of them. If this is what is meant by "odd", one wonders quite what qualities a newly born infant must possess to be recognised as an obvious "James". I, too, share my first name with these two American former policemen. Yet quite what characteristics we must have had in common to have all been lucky enough to share one of the most widespread names for male babies born in largely Christian countries, I do not know. By naming me "James", perhaps my parents expected me, too, to grow up to be a stocky American deputy sheriff, patriotic and with a burning need to go to church every week? If this was the case they must have been damned disappointed to get a scrawny Welsh political theorist, humanist and atheist to the core.

Whether those two babies were called John, Paul, George or Ringo should hold absolutely no interest for a geneticist, because babies' names are *externally* imposed. These twins did not select their names. For a geneticist not to realise this is extraordinary. This error is shocking enough, but to a Darwinian it actually pales into insignificance when compared with what came next.

It "was uncanny that each man had married and divorced a woman named Linda, then married a woman named Betty". One of the truly frightening things about many geneticists (as well as the genetics-quoting psychologists who tend to conduct twin studies) is their remarkable lack of familiarity with evolution. The American molecular biologist Robert Pollack has been trying to point this out for at least a decade. See, for example, his *Signs of Life* which decries the lack of understanding of the "history" of genes within his own discipline. "Trained as scientists but thinking like historians, biologists studying evolution have always embraced the contingent aspects of current and past life", Pollack notes. So far, he tells us, this kind of thinking has failed to inform the agenda of molecular biology. Although things are now starting to change, "[m]y generation of molecular biologists will likely continue as before". And yet evolutionary knowledge is critical for most areas of genetics. As the celebrated geneticist Theodosius Dobzhansky wrote in 1973 after almost four decades of leadership in the field: "Nothing in biology makes sense except in the light of evolution".

Now, if you are trying to study a mutant gene that has serious deleterious effects on the body, you can just about study it in isolation. But the problem with something like a gene for behaviour is that you must (a) know the normal behaviour coded into a typical animal in the first place, and (b) have a satisfactory explanation for how any genetic code you are looking for arose and was selected. Genes are only in bodies because of evolution. This is all evolution actually is: genes plus time. To ignore the time factor, and its considerable implications, is rather dangerous. So what has this to do with our two Lindas and our two Bettys?

Bouchard is suggesting that the twins were genetically coded to choose these women because of their names (because otherwise bringing up their names has no bearing). But since none of our character attributes has any connection whatsoever to our externally imposed names, there is simply no way for such a gene to have been selected. I have known some Lindas who are gentle and softly spoken, and others who could give Lizzy Borden (or Betty Borden as she'd be known in contemporary Massachusetts) a run for her money. Posit, if you so wish (and as evolutionary psychologists do), that genetics makes us look for partners who are tall, or attractive, or rich. But to suggest that genetics makes us look for partners that are called Billy-Bob or Betty, is, frankly, surreal.

Behavioural genetics all too often speaks to our gullibility and our rather poor mathematical knowledge. If you take any group, statistically you will find "uncanny" similarities between any two members if you look hard enough (and ignore all the bits that are not similar). Behavioural geneticists tell you "look, these two twins have first and second wives with the same respective name, and these two twins both chose the same model car". They tend to neglect to point out that the first set of twins chose different model cars, and the second set of twins have first and second wives who do not share respective names. When behavioural geneticists can produce independently collected data, carefully peer reviewed by people whose scientific training comes from other than the pages of the *Minneapolis Tribune*, and then produce facts that are genuinely statistically improbable, they will begin to be listened to a little more closely by social scientists.

Yet Thomas Bouchard is the most famous behavioural geneticist in America, and the two twins mentioned ("The Jim Twins" as they are known on the American interview circuit) are his most famous case. Not only are the common sense mistakes frightening, the lack of scientific rigour here is shocking. Science is done by setting up control groups and comparing experimental results against the control. Yet many identical twins both reared together and reared apart but later reunited actually define themselves by their relationship with one another, much more so than non-identical twins do, largely because of the attention factor. For example, identical twins take over an entire town in Ohio each August, hold festivals, and give each other awards. So there is a serious question mark over the value of studying two subjects who have already had a chance to interact, and a chance to demonstrate "uncanny" similarities to the *Minneapolis Tribune*.

Another methodological problem with behavioural genetics is the fact that some twins' researchers refuse to release the primary data on their subjects, claiming concerns over privacy (Twinsburg, Ohio, residents may feel a little surprised at this concern). Furthermore, some experiments become impossible to replicate - and replication is the cornerstone of scientific method - because there is a severely limited pool of monozygotic twins reared apart and later reunited (such social engineering is no longer performed), and the problems of coaching or fame-seeking make it dangerous to re-use subjects already interviewed. For those interested in understanding more about the issues here see James (2002) for a summary of the more common methodological problems with behavioural genetics.

The Times has a history of misunderstanding evolutionary biology, and Alfred Russel Wallace actually singled it out in his essay entitled "Creation by Law", first published in October 1868, and quoted from at the start of this section (I take my copy from part I, chapter VII of Wallace's *Natural Selection and Tropical Nature*). So when the above article was carried in *The Times*, I was not surprised. But when the BBC

Blessed of Nature

screened a three-part series in July 1999 around the "uncanny" similarities of twins I felt Wallace turning in his grave. Of course twins can seem spooky; natural selection strongly influences body form and so identical twins will necessarily tend to look similar-to-very-similar. And as we shall look at in the next chapter, genetics almost certainly does play an (as yet undetermined) part in certain attributes, such as intelligence: you do get slight variations in intelligence across other species and so *might* therefore expect some innate differences in human intelligence. Intelligence then goes on to play a part in ruling in and ruling out what opportunities one may get in life. Genetics may also have some input into certain tastes and mannerisms. But to say that differential genetics can, even partly, determine such anti-Darwinian concepts as moral behavioural difference is palpable tosh. And yet otherwise intelligent people seem to disengage their brains when confronted by twins who look identical, and seem to want to believe in "uncanny" similarities, no matter how strange the suggestions being made.

The BBC series under discussion was presented by Robert Winston, a professor of fertility medicine who had previously produced an award-winning series for the BBC on the ages of man. The final episode in that series, based around the last months of Herbie, a terminally ill man, and his adoring wife I found beautifully observed and very moving. Yet Winston's subsequent attempt to explain human life in his twins confessional, while sometimes good, was all too often ill-considered. This "major new BBC series" was trailered using The Jim Twins and their nomenclature gene. Winston, as easily as *The Times*, swallowed the idea that the names of The Jim Twins and their wives were "uncanny". Yet Winston is an intelligent man, and respected in the field of developmental medicine. His medical background should have given him at least a nodding acquaintance with evolutionary genetics. As far as I can see he (unlike a number of the actual twins researchers) is not noticeably political in this field. Yet can this senior figure in the British medical establishment really believe that evolution produced a creature that selects its mates, not by their brains, speed, effort or physique, but by their *names*? Names imposed on them by others, and which can have no connection to any other character attributes? Surely in the months it took Winston to make this series, it would have crossed his mind that something wasn't quite right?

People seem to willingly accept at the individual level what they would put a lot more thought into at the group level. Had the claims of behavioural geneticists been used to openly support a view of differences across groups (as they once were, and sometimes still are), Winston would probably have thought twice about what he was suggesting. A March 1997 U.S. Department of Justice report ("Lifetime Likelihood of Going to State or Federal Prison") found that while white American males have a 1 in 23 chance of incarceration, for Hispanic males it is 1 in 6, and for black American males it becomes greater than a 1 in 4 chance. In the United States many people may be credulous enough to believe (or, rather, want to believe) that this has something to do with differential biology, but Joe Public in Britain tends to join evolutionary psychology in rejecting such prejudiced nonsense. But when you bring it down to individuals, critical reason so often goes out the window, even when the summation of such individual "genetic" differences almost automatically begets the group prejudice. Faith is the enemy of reason, be that faith in the divine, or faith in the gene. Human credulity over twins research is appalling. Too little oversight of the original research occurs, all too often claims are subsequently disproved or withdrawn, and even where the work is good it tends to make mountains out of rather unimportant molehills. Yet bad twins research continues because of the public appetite for it. We are so desperate to give substance to our prejudices, our hopes and our fears, that we eagerly swallow gobbledegook about angels, E.S.P., and "uncanny" twins. There is much that twins research can meaningfully tell us. But for that we will need far better teachers, and far wiser audiences.

Oh, and for any of you still puzzling over the two Jameses "uncannily" both turning out to be policemen, please try to remember that being a policeman is one of the few attractive career options for working class white American males.

Unwelcome enthusiasts
You can ground no politics in Darwinism. The very term "politics" is derived from the Greek word *polis* which means "city". Cities are built by humans co-operating relatively peacefully in their thousands and millions. Not by groups of one hundred living in untrusting relationships characterised by unbelievable levels of violence. **Not by groups living, in other words, as apes.** So we can ground neither Wilson's politics, nor Marxism, nor anyone else's politics, in gene-

selectionist Darwinism. From Ayn Rand-ian right wing canons to Marxist left wing doctrine going via humanist liberalism on the way – all are products of cultural manipulation, not genetic manipulation.

Yet genetics and Darwinism will continue to be misused in the service of political, or nationalistic, or even religious, ends until we start to be a little more discerning; until we start to accept that we are products of evolution by natural selection. Biology needs to police itself, yes, but this can never be enough. As Karl Popper warned, to trust to the impartiality of the scientist is dangerous, and sometimes even the impartiality of the scientific process itself can be legitimately questioned. We all have a duty to think a little more carefully about the lines we are being fed. Too many people have suffered because of bad science. Too many continue to suffer because of bad science.

7

Magnificent exceptions?

Life and all its glories are thus united under a single perspective, but some people find this vision hateful, barren, odious. They want to cry out against it, and above all, they want to be magnificent exceptions to it.

Daniel Dennett – Darwin's Dangerous Idea

Most of the hypotheses put forward in the last 25 years of heretical human Darwinism are easily dismissed when one considers man as - genetically - just another ape. Yet there are some of these conjectures that are not so obviously wrong, particularly those that come from the new generation "evolutionary psychology". In this chapter we consider one or two that seem to make the grade - until you look more closely still. We shall also consider the cases where evolutionary theory can currently do little more than provide important insights.

Grief
Robert Wright was a science journalist who read about studies that

Magnificent exceptions?

showed that human parents grieve more over the death of an adolescent than they do over the death of an infant. Such studies suggested that there was a Darwinian reason for this, as adolescents were "assets on the brink of bringing rewards". After the many years of effort expended in raising it, the adolescent is on the brink of fulfilling its purpose by reproducing and thus passing on the parents' genes. Few resources have yet been invested in the infant, so it is natural that the parents should grieve less. "The evidence so far is that grief does comply exquisitely with Darwinian expectations", Wright wrote in his bestselling book *The Moral Animal*. Logical, hey? Well, yes, except for the slightly important point that grief exists nowhere in the non-human world.

Helena Cronin was an evolutionary psychologist with enough knowledge of natural world behaviour to begin to be slightly worried by the hypothesis. All people will pick out a photograph of a grieving adult as the normal expression of someone who has just lost their child, she noted. "[I]n any culture you go to they will all pick out the same. So it seems that it is indeed innate and part of our evolutionary repertoire." Nevertheless, Cronin was to admit that: "I don't really understand why we have grief though. It doesn't seem to me to be advantageous to the individual who feels grief, and it's not obvious to me why we do" (speaking on BBC Radio 4's "Basic Instincts". See chapter 3 for details). And that was actually the point. Grief will not arise under natural selection because it is (from an evolutionary point of view) pointless and inefficient. And natural selection is simply an efficiency selection process; if something is not advantageous to the individual it will not become a part of the natural world. In the battle for gene survival wasting energy on gene vehicles that have not fulfilled, or will not fulfil, their purpose is counterproductive. So grieving for them after the event is even more senseless.

Chimpanzees can certainly feel frustration, and a chimp may even drag around the carcass of its dead infant for some hours. However, such temporary frustration does not come close to paralleling heartfelt human grief, while suggestions of human grief as genetically coded, as adaptive, are just plain wrong, for the reasons given above. Grief *had* not evolved because grief *would* not evolve.

Yet poorly understood gene-selectionism is the story of evolutionary psychology. But while some of the stories are relatively easy to dismiss, others require a little more thought.

Magnificent exceptions?

Human infanticide: A natural kind?
One claim of evolutionary psychologists is that the psychologists Martin Daly and Margo Wilson effectively "proved" that step-parents kill stepchildren for "Darwinian" (survival of the genes) reasons. Statistically such children are up to 100 times more at risk of fatal abuse than children living with their natural parents. It is the one study that is *always* quoted, and always claimed as being "intellectually rigorous". At first sight the hypothesis seems to have something to it. Step-parenting does not occur in the natural world. The young male offspring of a female from another group is torn to shreds by chimpanzees because Darwinian "pitiless indifference" will not allow an ape to raise another's child. Isn't this an evolutionary psychology theory that *finally* accepts that humans are also born as selfish, cannibalistic, infanticidal apes? Well, no.

Where Daly and Wilson collapse is their Darwinian logic. Let us take a closer look at their study.

Remember that evolutionary psychology is, in general, about *universal* features of the mind. Now Daly and Wilson are arguing that infanticide is being genetically determined in these homicidal step-parents, exactly as it is in chimpanzees. So far the hypothesis is plausible. So it is a universal feature of all humans? We are *all* genetically coded for it? When I first came across the suggestion that evolutionary psychologists had proved a "Darwinian" rationale for the evil stepmother fairy tale, I wondered if this was some weird variant of the "Confessor" Complex. A coded warning to their future spouses by evolutionary psychologists, perhaps: "Dear, please don't leave me alone with your children. As an unreformed ape I wouldn't trust myself not to make a casserole out of them". But, oh no, this is not what evolutionary psychologists mean at all. We might *all* grieve for our lost child, but cannibalism or the infanticide of a stepchild are not quite such a normal "part of our evolutionary repertoire" according to evolutionary psychology. Only *certain* humans will ever act in this way. This is that *other* definition of "universal".

Scientific American sent John Horgan to research evolutionary psychology in 1995. What Margo Wilson told Horgan at their interview I find nonsensical. She told him that adopted children do not need to be similarly studied because they are no more likely to suffer fatal abuse than natural children are. But Daly and Wilson's argument is that step-

Magnificent exceptions?

parents kill their stepchildren because they have no genetic stake in their future, and indeed the children are genetic undesirables from the point of view of the step-parent. Yet stepchildren still have one parent around interested in their genetic welfare. In contrast, adopted children have no parent around interested in their genetic welfare. Using the Daly / Wilson reasoning, adopted children should actually be at a *higher* risk of fatal abuse than stepchildren should if, indeed, we are dealing with a hangover from our Darwinian past. In their book *Homicide* Daly and Wilson suggest that adopted children, unlike stepchildren, are at no additional risk over biological children because the adopting couples are often childless couples, "strongly motivated to simulate a natural family experience and who have been carefully screened by adoption agencies". Now the second point should make no difference if we are truly dealing with a "part of our evolutionary repertoire" and an equal basic genetic programming. But perhaps some of us are more equal than others? As I showed in "Unnatural Selection", evolutionary psychologists just cannot give up this desire to split us into the genetic elect versus the biological reprobate. But let us ignore their second point, and turn to the more fundamental issue of the first point.

People adopt because biologically they are "strongly motivated to simulate a natural family experience". What we see now is that at the heart of the Daly / Wilson hypothesis is this heretical conception of man as the "decent" ape. As the human sociobiologist David Barash wrote in 1979: "[o]ur evolution leaves us with a rather 'open program' that enables us to adopt children comfortably". Oh, what a *nice* little ape we are. There is just one *tiny* problem. This is anti-Darwinism of the highest order. Adoption of genetically unrelated children is just about the most devastatingly stupid "strategy" you can think of from the point of view of natural selection. As Richard Dawkins writes: "[t]o use the language of Maynard Smith, the altruistic adoption 'strategy' is not an evolutionarily stable strategy". And because it would be, not just a pointless "strategy", but a counter-productive "strategy", it cannot have evolved. This "tendency" would automatically breed itself out of a population. "[W]hen, as frequently happens, people challenge Darwinians to 'explain' the love of a woman for her adopted child, say, it is often sensible not to accept the challenge. Civilized human behavior has about as much connection with natural selection as does the behavior of a circus bear on a unicycle", Dawkins and Mark Ridley wrote in 1981.

Magnificent exceptions?

Now this is not to say that instances of adoption cannot occasionally happen in the non-human world, but we need to be very clear about what is and is not going on, and that such instances cannot be adaptive, can only be maladaptive. Cuckoos force adoptions on other birds, force maladaptive behaviour on those other birds, and we have already dealt with this pattern. Very occasionally primatologists spot a chimp whose infant has died taking on a non-closely related infant. In *Adaptation and Natural Selection* Williams pointed out that misplaced parental care can occur, often only briefly, when certain animals have their reproduction frustrated. Selection pressures maintain a certain pattern of parental behaviour, he wrote, but with a less-than-perfect system of timing mechanisms for regulating this behaviour. A few years ago a small child fell into the enclosure of a mother gorilla who the previous year had been given human dolls to play with pending the birth of her baby (to try to stop her rejecting it, such rejection being common in gorillas). Onlookers screamed, but, far from killing the child, the gorilla carried the child safely to her keepers. "I can't say that it was a typical response", the curator of the gorilla department noted dryly (Allen-Mills [1996]). Maladaptive mistakes can and do occur, even in infanticidal chimpanzees and gorillas, but this is so very far from evolutionary psychology and its morality-as-a-biological-adaptation creed; so very far from man as the "decent" ape. We are born cannibal and, as Dawkins says, it really is not sensible to try to explain human adoption with an evolutionary story.

Daly and Wilson have said that children in stepfamilies are more likely to be abused because there is less likelihood of love as there is no genetic bond. But what they have missed is the point that love (like heartfelt grief) does not exist in nature. If children in families - be they natural families or stepfamilies or adoptive families - are being loved, we cannot look to genetics for an explanation.

So could there be another explanation for why step-parents are statistically more likely to kill stepchildren? One that also explains why adopted children are no more at risk than biological children (and, crucially, also starts from the Darwinian knowledge that we are *all* born *selfish*, not altruistic, apes)? As Dennett has written, simply taking statistics (such as higher fatal abuse rates for stepchildren) and using them to construct a "Darwinian" rationale is a dangerous business. It can easily lead to bad science unless it is done carefully and with full

Magnificent exceptions?

understanding of evolutionary genetics (such as the slightly important point that we are *not* born genetically capable of adoption).

This illustrates the fundamental obstacle – not insuperable, but much larger than is commonly acknowledged – to inference in human sociobiology: showing that a particular type of human behavior is ubiquitous or nearly ubiquitous in widely separated human cultures goes **no way at all** *towards showing that there is a genetic predisposition for that particular behavior.*

Daniel Dennett – Darwin's Dangerous Idea

Gould, too, stresses this point. "In any case, even if we can compile a list of behavioral traits shared by humans and our closest primate relatives, this does not make a good case for common genetic control". Similar results need not imply similar causes, he says. "[I]n fact, evolutionists are so keenly aware of this problem that they have developed a terminology to express it. Similar features due to common genetic ancestry are 'homologous'; similarities due to common function, but with different evolutionary histories, are 'analogous'."

So what would be a better explanation? Well, let's think about it. Do you love your children? If your child were to die tomorrow, would you feel no grief? Would your only thought - after frustratedly dragging its carcass around for several hours - be: "Better start working on another one"? I would hazard that this would not be your reaction. And yet this is the correct *Darwinian* reaction. So your basic attitudes are not set by genetics – and as John Maynard Smith tells us: "[f]or a geneticist, all variance which is not genetic is, by definition, 'environmental'".

I hold the frighteningly British opinion that you often seem to find the wisest minds in North America writing sit-coms. In the wonderful sit-com *Frasier*, Frasier told mother-to-be Roz that, watching his sleeping small son one night, he realised that "you not only love your children, but you fall in love with them, too". Human child-parent relationships are about far more than genetics. They are about love and empathy, concepts that do not exist in nature. *This* is the reason we feel grief at the death of a child, and *this* is the reason that we view our children not as "gene vehicles" but as people. Evolutionary psychologists make an important point when they say that by studying such statistics as higher fatal abuse rates for stepchildren we can learn lessons. But such lessons have *far* more to tell us about culture than they do about genetics.

Magnificent exceptions?

Stepchildren *are* more at risk of fatal abuse, and by accepting this we might begin to reduce the risks to such children. But evolutionary psychologists grasp totally the wrong end of the Darwinian stick when they start by asking why some step-parents do not grieve at the death of an unloved stepchild. This may indeed have no genetic component anyway, as Dennett and Gould warn us, and as we consider again below. But the question that *is* of overriding importance to a Darwinian is why most step-parents *do* grieve at the death of a loved stepchild - and why any of us grieve at the death of a loved *natural* child in the first place. This is humanity riding roughshod over genetics. This is being human.

Cultural "manipulation", the answer of the biologists, means that human kindness, such as adoption, is made possible by culture (acting on our very large brains) twisting us free from our genetic inheritance and establishing new behaviour patterns. But the same is true for human nastiness. Our individually unique cultures, our local "environments", teach us who to feel empathy towards, and who not to feel empathy towards. Biological parenting involves adults raising another being that is initially totally dependent on them, that can be raised and nurtured in their image, and that (if they get it half-way right) will spend the first ten years of its life worshipping them. Adoption, too, involves such anti-Darwinian empathy. It is often about adults feeling empathy towards, either orphaned children *in toto*, or a particular lonely child, and then choosing a life with a child. Step-parenting is quite different. What happens here is that an adult chooses a life with another adult, but then suddenly (and with no time to get used to the dramatic lifestyle changes) finds that a child also comes as part of the package. A demanding child often with already-formed opinions and character who may not want to watch 20 hours of *American Gladiators* every day. A child who is not looking for another parent and will make that quite obvious. What then happens is that tensions and pressures build up, often in homes where there is no easy way to find space away from each other. And usually in homes where (flying directly in the face of evolution) the new parent is happily allowed by the biological parent to come before its child.

Stepchildren are appallingly abused every day, and far more commonly than biological or adopted children are. But these children are no more abused for "Darwinian" reasons than kids enjoy playing in mud because, three and a half billion years ago, their ancestors were

Magnificent exceptions?

pond slime. Evolutionary psychology, with its implicit suggestion that, since it's genetic, helping such families undergoing stressful changes may be pointless, is music to the ears of those who don't wish to embark on the often long and painful process of actually reducing such statistics. That, of course, would require far too much effort. It takes imagination, not wisdom, to make up Darwinian stories. It's a fairly universal reaction for adults to feel heightened awareness when they sense a small child about to dash out into a busy road. But something tells me that you would rightly feel affronted if anyone suggested that your attentiveness was nothing more than your infanticidal, cannibalistic genes taking a visceral interest in the potential extermination of a future genetic rival.

The mark of a true Darwinian is the ability to discriminate between good and bad Darwinian theorising. Darwinism is about *indifference*, not love and empathy, and not contempt and hate.

Attention-Deficit Hyperactivity Disorder, or The Preservation of Favoured Parents in the Struggle to Pass the Buck

Christianity without the tears – that's what soma is.

Aldous Huxley – Brave New World

"Christianity without the tears – that's what *soma* is", explained Mustapha Mond, one of the ten World Controllers in Aldous Huxley's *Brave New World* (Aldous, incidently, was a grandson of Tom Huxley). "Soma" was the drug used by the World State to control people in the "production line society" of Huxley's famous novel. "Soma" did away with the need for great effort and years of moral training. Yet Huxley's satirical prediction is already true. Welcome to the "production line society".

There is much talk in the United States about "Attention-Deficit Hyperactivity Disorder" (or "ADHD") and the equally terrifying "Attention-Deficit Disorder" ("ADD"). Horror! American schoolchildren can no longer concentrate! The rate of ADHD/ADD diagnosis is reaching epidemic proportions! Thank God for our pharmaceutical giants, ready to charge bravely into the breach for us and solve all our problems! The solution to ADHD children is to dope them up to their eyeballs on behaviour-modifying drugs such as Ritalin. Similarly, the antidepressant Prozac is now prescribed in the States for everything from shyness to

Magnificent exceptions?

poor school performance. Ritalin and Prozac, saviours of the good old U.S. of A. And this epidemic has now reached out its terrifying fingers to touch European minds. Almost 200,000 British children are today regularly prescribed Ritalin. Prozac too is dispensed with abandon to adults and children alike.

The American philosopher Daniel Robinson commented wryly on "ADHD" when he appeared on Melvyn Bragg's BBC Radio 4 series *In Our Time* in December 1998. ADHD, he said, was one of America's national epidemics. "We usually have two or three at a time in the States". An old school friend of his had called him up because his grandson had just been diagnosed as having ADHD. Before they resorted to the doctors' advice of powerful drugs, the grandfather had sought Robinson's view on this child who had been clinically diagnosed as "unable to concentrate". "Does he like sports?" Robinson asked. "Loves them", was the answer. "When his favourite sport is on the television, does he stay glued to the set?" "Can't drag him away", was the response. "And he can't concentrate, hey?" The geneticist Steve Jones is even more sceptical of psychologists and their need to come up with rather meaningless acronyms. In *The Language of the Genes* Jones finishes his chapter on the continuing abuses of human genetics by commenting: "Now psychologists have invented 'attention-deficit disorder'; something intrinsic to the child and coded in the genes. Psychology's obsessive need to dissect biology from experience is alive and well".

There is a multitude of reasons for why so many American children will not now concentrate in school. Reasons that are complex and interwoven, and allow for no easy answers. Sometimes they have their roots in poor parenting, and sometimes they don't. Peer pressure, home life, expectations, distractions, interest, ability. They all play a part. But there is one reason above all for why these children have a problem. Our children suffer because we are not smarter.

Anglo-American culture has always tolerated a particular ideological blindness when it comes to raising children. We like to feel that raising a child is somehow both innate and easy. That it doesn't take much effort, and we know how to do it anyway. Then in the 1980s we began to see the publication of works which again claimed that genes, not upbringing, were the critical factor in child development, and that parents needn't feel guilty about placing their careers ahead of their

Magnificent exceptions?

children because biology had already written their children's future. In the late 1990s there was an influential modification to this doctrine in the work of Judith Rich Harris, who thrilled evolutionary psychologists like Steven Pinker by producing a book cataloguing the considerable influence of peers on children's behaviour. When I first heard about Harris's book (*The Nurture Assumption: Why Children Turn Out The Way They Do*) I couldn't quite understand what the excitement was about. Of course peer pressure is important in child development; in *The Descent of Man* Darwin singled out positive and negative peer pressure as to him perhaps the most important factor in the setting of moral norms (and subsequent moral development and control) in all age groups. I remember with a shudder the horror of peer pressure at school, and the powerful external and internal requirements to conform. Unfortunately some psychologists have tried in the past to claim that parental influence is all that matters in child development (partly, I believe, through modern intellectual laziness, as many academics want simple answers to their questions, not complex interacting factors). However Harris's good work is largely undone by then trying to deny that parents have any real influence on how their children turn out - unsurprisingly the book was an instant bestseller among busy American parents. For an example of the anger directed at this claim by some members of the psychological community see the comments of clinical psychologist Oliver James in his *They F*** You Up* (James [2002]). I am not a psychologist, so cannot pass judgement on James' comments, but as an evolutionary theorist I can tell you that in young children especially, parental nurture must be absolutely critical. By the time a child even attends playschool a grossly immoral ape-child has been radically re-written. Given how grotty little kids can be when you get them together we can be reasonably sure it was the adults around them that inculcated the "now, don't eat each other" ethic, rather than their peers. Harris's work is unfortunately based on the "(Mother) Nature Assumption" (that we are born moral apes), and she appears to place a surprising level of confidence in the claims of behavioural geneticists.

This is the resurgent American media view of child-raising, it seems. Parents are powerless, and if your child causes you problems simply blame its peers, blame your partner's genes, or, if all else fails, invoke genetic mutation or ADHD (most psychologists will be only too willing to co-operate for a fee). Raising a human child is complex, messy and time consuming. Unfortunately all too many parents don't want to acknowledge this,[24] or have been led to believe that child raising is innate. Knowing how to raise a chimpanzee is innate; knowing how to raise a human must be learned.

One factor in the ADHD "epidemic" that seems to have come out of research over the past couple of years is the crucial importance of play to children. Play is not only important as a release of tension. It is also during play that children learn to socialise, and to work in groups. They learn co-operation and that they cannot always expect instant gratification. And yet many American schools, desperate to climb those academic league tables, are starting to cut playtime, sometimes cutting it out altogether. We really are our own worst enemy, aren't we? Despite what political ideologues want you to believe, no child is immune from developing ADHD. It has its roots in a changing society, and human myopia. Oh, joy! One day, if we *really* try hard enough, 100% of our children could be on Ritalin.

As Darwin understood, to make a well behaved member of a community requires considerable time and effort. Until, that is, the development of Prozac and Ritalin.

There are, then, two routes we can go down. We can invest in our children, and ensure that our politicians are informed enough to realise, for example, the damage that cutting playtime can do. Or we can leave it up to the drug companies to make our children for us. Unfortunately, as the experience in America has shown, the problem with the latter course of action is that the underlying problems will just keep getting worse until we start to get smarter. Whereas children haven't changed biologically in one hundred thousand years, human societies have changed substantially over the last few decades. All children are still born selfish monkeys. But where once we captured their minds with

---

[24] Steven Pinker wrote the Foreword to Harris's *The Nurture Assumption*, and his own work *The Blank Slate* is a more recent example of this genre of denying parents have any real control over their children's progress. Worryingly, Bugental *et al.* (1989) found that parents who actually believe that they have little control over their children's development and behaviour are more likely to maltreat them. Furthermore, they noted that when an at-risk child is paired with an adult with low perceived control "a dysfunctional match is created" (Bugental *et al.* [1989], p. 538).

stories, myths and rituals, now we cannot afford the time, or find that we are not bright enough to compete successfully with their other distractions. So, instead, we keep them doped up to the eyeballs on pharmaceuticals that pacify them and make them easier to control.
Welcome to the brave new world[25].

ADHD revisited - a scientist's guide

I am not trying to downplay the seriousness for many parents with kids diagnosed as "having" ADHD, or denying that "it" can exist. These kids are often very difficult to handle. But almost always these children *became* difficult to handle. In the great majority of diagnosed cases of "ADHD" we can be sure those children were not born any more difficult to handle than other more "normal" children were. And we can be pretty sure of this because all children are born genetically programmed to behave in ways that make Genghis Khan look like a shrinking violet.

Yet ADHD is not grief, and it is not infanticide. Evolutionary knowledge makes a mockery of a genetic component to grief, just as it raises serious questions over a genetic component to human infanticide. But while the great majority of children diagnosed as having ADHD were born no more difficult to handle than other children, there do seem to be occasions when certain ADHD children were innately more susceptible. Many diagnoses of ADHD, and most diagnoses of ADHD as something a child was "born with", are a reaction to moral panic in overly credulous publics, as the Robinson anecdote nicely suggests, and as basic evolutionary knowledge confirms. All children are born with a grossly immoral genetic code that has to be overcome by training, and in some cases the training is better than in other cases. But this does not mean that there may not *also* be some genetic or chemical component to why a small number of children are more (hyper)active than others.

---

[25] After drafting this section I came across an article entitled "Drugged-Out Toddlers" in *Newsweek* from 6 March 2000. It reported on a study published in *The Journal of the American Medical Association* which showed that the use of certain psychotropic drugs in 2- to 4-year-olds doubled or even tripled between 1991 and 1995. "America's tiniest citizens, some still in diapers, are now the newest members of the Ritalin and Prozac nation. ... [E]xperts say frustrated parents, agitated day-care workers and 10-minute pediatric visits all contribute to quick fixes for emotional and behavioral problems" (Kalb [2000]).

Magnificent exceptions?

Most hyperactivity will have a cultural root (or at least most "hyperactivity" only becomes noticeable because certain children do not get a chance to learn the normal self-controls on behaviour that most children will have a chance to learn without thinking). For all I know, some pill-pusher might today classify me as hyperactive if I had to re-live my childhood again. Since childhood I have always been on the go, my mind continually racing (although maybe not always in the correct gear), and my body not far behind. Yet I learned when it was appropriate to express such energy, and when it was not. And maybe there is some innate biological input to my excess of energy, just as some children diagnosed with ADHD may have some biological or genetic handicap.

The proportion of ADD/ADHD cases that have a real inherited susceptibility is likely, however, to be rather small. Ian Robertson is a leading name in the field of neuropsychology and neurorehabilitation. In his 1999 book *Mind Sculpture* he sought to explain how parent-infant interaction physically *builds* neuronal connections within the child's (and, to a lesser extent, the parent's) brain. Reading with your child, talking or playing with him or her, improves your child's memory, intelligence and social skills, actually building pathways within the brain. "The ability to pay attention - to concentrate on something - is one thing that we teach our children every moment of every day." Attention-deficit disorder, Robertson notes, is a very commonly diagnosed problem in the United States, Australia and The Netherlands, though it is less frequently diagnosed in Britain. Some of these children really do have differences in their brains from birth that make it harder for them to hold their attention to any one thing for more than a few minutes, he says, "but many more simply have not had their brains' attention systems coached and trained into working well. ... But just as sculpting the brain requires effort, attention and practice by the child, so it needs the same things from busy parents who have their own worries and preoccupations".

Many neural scientists are understandably horrified by the increasing tendency in our fraught Western cultures to try to deny that parents play a vital role in helping their children to develop mentally. Adult conscience-appeasement is a very dangerous game to play. The insight of those like Robertson who study developing brains, congenitally abnormal brains, and accidentally damaged brains, is largely confirmed by evolutionary theory. His insistence that "many more" children

Magnificent exceptions?

develop ADHD through poor stimulation than ever develop it through brain damage or some innate tendency is exactly what Darwinism implies to theorists like Steve Jones. Where we are talking about a child with an inability to concentrate from birth, or from a subsequent accident, we are talking about an abnormal brain. But such brains are rare in nature. They are rare in chimpanzees and other primates. And they are rare in humans, as brain abnormalities like autism or retardation suggest, and as statistical gene mutation explains. If you are invoking a new brain disorder where even the symptoms have rarely been catalogued in medical or social histories, and then are suggesting that this "innate" disorder can be applied across great percentages of the population, evolutionary genetics will catch you out. Phenotypic variability among genotypes may be increased in abnormal environments, but this too cautions us that we may be raising more and more children in "abnormal" environments. And if we continually strive to make these environments more and more abnormal there *will* come a point when no child will be free from ADHD. The attention deficit often seems to be more in our politicians, our parents and our psychologists than in our children.

Almost all children with ADHD will benefit from powerful drugs. But the great majority of those children should never have needed drugs in the first place. Unfortunately few psychologists have the inclination to say this. After all, if a drug can mask any problem, whether its origin be genetic or cultural, it takes a very strong-willed medic to stand out from the crowd and ask "Why?" before he reaches for his prescription pad. Yet establishing the causes of ADHD in every particular case is crucial (and Dan Robinson would probably like to point out that actually making intelligent, rather than bandwagon, diagnoses in the first place might help).

Until we know the causes in each specific case we cannot be said to be acting in the child's true interest. Handing out pills to children without seeking at the same time to find out why they first began to need those pills may be a great relief to both busy parents and timorous medical practitioners, but it does little for the child. Our kids deserve good parents, not just good pharmacies.

Oh, and before we leave this chapter we will ask the *most important question of all* - if some children do have some slight innate biological difference from the norm in this instance: "So what?"

Magnificent exceptions?

Intelligence, schizophrenia, alcoholism, depression and homosexuality

Now we shall look at the genetic component to intelligence, schizophrenia, alcoholism, depression and homosexuality. This section is easy. *I don't know*. I don't know what the genetic components are, because these issues largely fall outside my specialism. I am sorry that I could not be a little more definitive on what the genetic component may be in a small number of cases of ADHD as discussed above, and I am sorry that I cannot be definitive in regard to these issues. I am an altruism theorist; this means I study the "gross immorality" of nature. I can tell you about evolutionary morality - there is none. I can tell you that you were born a cannibal. That knowledge has important implications for many other areas of study, but there is not always a total correlation, particularly where we come to areas where differences *are* found across other species. There must be some genetic input to the variations in human intelligence, although human culture is immensely powerful at raising standards and lowering levels of intelligence. As the geneticist Steve Jones points out, if you want a child with a high IQ, forget spending thousands of pounds on genetic screening and just send him to Eton. Early stimulation and then a good education are absolutely crucial in building high IQ in almost any child. But as to the actual genetic components to human intelligence, that Darwinian theory cannot yet tell you. It can, however, remind you of many important lessons.

Let us briefly consider IQ as an indicator of intelligence (and ignore serious questions of how good an indicator it is and many other associated problems). The evolutionary biologist Sir Peter Medawar, in his 1977 article "Unnatural Science", notes for us both the extent to which some scientists will go in order to prove a pet prejudice, and the poverty of understanding of this complex issue even among IQ researchers. To set the scene he begins by discussing the notorious case of the late Sir Cyril Burt, the British educational psychologist, twins researcher and professor at UCL. As Medawar notes, it was a team of investigative journalists at the *Sunday Times*, led by the paper's medical correspondent Dr. Oliver Gillie, who began an investigation which ultimately "questioned the probity of Burt's entire work", and concluded that many of those who had supposedly supplied Burt with data did not, in fact, exist. The investigations followed earlier criticisms

Magnificent exceptions?

of inconsistencies in the data by the geneticist Leon Kamin. Burt's work, which had concluded that there was such a thing as general (not specific) intelligence and that it was largely fixed by age eleven, was the single greatest driving force behind the "eleven-plus" examination in Britain. This was an examination that effectively determined forever many children's position in life on the basis of a test taken at age eleven. The eleven-plus hurdle has now been abandoned in its original form and intention, and even where it still exists it has been redesigned so as to avoid restricting the opportunities of those who fail (although I have been asked to note the quasi-eleven-plus today operated by both local authorities and oversubscribed private schools). The original eleven-plus was driven by the good intentions of politicians, but ultimately ended up depriving millions of youngsters of a possibly much more successful future because of the scientific naivety of those very same politicians (and the public at large). But why did Burt act deviously, asks Medawar. "Villainy is not explanation enough: Burt probably thought of himself as the evangel of a Great New Truth." Evangelists, in all their forms, can do such immense harm, even where they may not intend to.

Those keen to understand the genetic component to IQ must be cautious not only of those who deliberately skew data to back a pet prejudice (be that the higher or lower intelligence of a certain group, class or race), but they must be careful that they even understand their own question in the first place. Variations in intelligence are partly inherited, but establishing the heritable component is extremely difficult because culture is so changeable, so powerful and so ubiquitous. "Human" is, after all, a cultural construct. As Medawar notes, the really important question is whether or not it is possible to attach exact percentage figures to the contributions of nature and nurture. "In my opinion it is *not* possible to do so, for reasons that seem to be beyond the comprehension of IQ psychologists, though they were made clear enough by J.B.S. Haldane and Lancelot Hogben on more than one occasion, and have been made clear since by a number of the world's foremost geneticists." The reason, he says, "which *is*, admittedly, a difficult one to grasp", is that the "contribution of nature is a function of nurture and of nurture a function of nature".

And on that note I shall leave intelligence to those far more qualified to speak on it and turn instead to another puzzle. In the last chapter we briefly looked at some reservations raised over one of the latest "gay gene" claims. Beyond this, though, my answer to the genetic component for homosexuality is that I just don't know. There have certainly been a number of studies claiming a strong genetic component which have subsequently been discredited or withdrawn. There is certainly great ideology driving a number of the researchers and their backers. But this does not rule out there being a sizeable genetic component to homosexual behaviour. As Williams told us in an earlier chapter, homosexual behaviour is certainly "common" in nature. It exists in all those other grossly immoral primates. And there is no necessary linkage that the culture that overwrites our cannibalism genes must necessarily overwrite all our biological sexual appetites. But this is a long way from saying that we are even touching on the truth yet. One thing that surprises me is that human sociobiologists and evolutionary psychologists haven't posited *more* theories on sexuality. What I mean is that I just can't help feeling that we should perhaps take more note of how common bisexual behaviour is in nature, and that such behaviour is *species-wide*. There are a number of suggested reasons (some fanciful, some very well evidenced) for homosexual behaviour in vertebrates, ranging from group cohesion, conflict resolution and bonding through to practice. One of the most amusing and illuminating examples is bonobo behaviour.

Our closest living relatives are the chimp and the bonobo (the latter found only in Zaire). Bonobos are fascinating because they use sex as a primary method of maintaining their small group bonds. Bonobos tend to live in groups of only a few dozen, while chimps can live in slightly larger groups. But whereas chimps maintain intra-group bonds by mutual grooming, bonobos also add lots of sexual play. As Dawkins put it in *Unweaving the Rainbow*: "Where we might shake hands, they copulate". Age and sex do not matter to bonobos; youngsters copulate together, adult bonobos copulate with youngsters, males perform sexual services for other males, and females perform services for other females. Copulation is so common and ritualised that half the time partners are hardly even paying attention. Yet what is even more interesting than this bonobo "attention-deficit disorder" is that evolutionary psychologists never suggest that *this* sexual genetic coding might exist in man too.

Bonobos are genetically programmed for such bisexual behaviour (some primatologists prefer the term "ambisexual"). Many other species

Magnificent exceptions?

of vertebrates are also known to be similarly coded (although frequency of homosexual coupling may not be so common because groups have different ways of bonding). Nevertheless, frequent or not, the genetic code for bisexuality is still there, perhaps to be expressed according to group conditioning. So why no evolutionary psychology hypothesis that we *all* carry the "genes" for bisexuality, but environment tends to determine who will have the opportunity to express them? Hmmm? What do you think? Keen on the hypothesis that every bigoted conservative actually carries in his or her DNA the coding for swinging both ways? What an extension to Hamer's "gay gene" claim: homosexuality isn't *on* the X chromosome, homosexuality *is* the X chromosome!

The above is posited not as a deeply pondered question, or even as necessarily a reasonable suggestion, but basically to make certain more homophobic behavioural geneticists apoplectic. I do, though, wonder whether we haven't even begun to learn from the sexuality found in the natural world, and that if we did the people who are so keen to posit genes for homosexuality, as well as those who wish to regard it as unnatural and immoral, might be a little more inclined to keep quiet.

(Any answer to the *expressed* genetic component to human homosexuality must determine first whether natural world homosexual behaviour and human homosexual behaviour are ever the same thing. Yes, no, maybe. It is a similar point to that already considered in regard to human infanticide. In the final section of this chapter we shall look at the implications when we make the huge - and I believe incorrect - assumption, á la Daly, that those rare instances of human infanticide can actually sometimes be equated to widespread natural world infanticide. A similar point could be made for human rape; first we must ask does it really have any connection to natural world genes-into-the-next-generation rape? *All* male primates seem to be genetically coded for rape, simply because they are, well, male primates. But this does not mean that those rare instances of human rape are necessarily still - sometimes or even ever - genetic in origin. Just as human virtue has no genetic component, so this often seems true of human nastiness. The more important point for social scientists to understand, of course, and as per the last section of this chapter, is that *even if* there is some genetic linkage, our *differences* in behaviour are cultural because we *all* carry the same grossly immoral natural world genetic code. We are all born infanticides, but - assuming human infanticide can sometimes be equated to natural world infanticide - cultural conditioning will then determine who displays such behaviour. It is culture that determines how we live our days. There but for the grace of a chance upbringing.)

Yet while I cannot begin to tell you the genetic component to human homosexuality, I *can* tell you that homophobia is 100% culturally conditioned. It is learned from religions and other ideologies, taught by men like the prophets we shall meet in chapter 9. It is not just that homosexual behaviour is so common in the natural world, it is also because nature is still about indifference. All animals - except man - couldn't give a hoot about another animal's sexuality, except perhaps for that brief period when they are mounting something ("brief" because although sex tends to be *very* common it also tends to be *very* short). To care about another person's sexuality tends to require a bigot in a robe somewhere along the line.

A future for evolutionary psychology

I don't want you to think that I do not see an important role for evolutionary psychology. Evolution cannot explain *love* (as we saw in an earlier chapter, and will consider again later), but it may be able to explain much human *lust*. Under consideration by zoologists is the plausible-sounding idea that in nature animals may have been positively selected for attraction to partners with symmetrical physical features, as this could be a useful visual indicator of healthy genetics. If symmetry is indeed a better-than-average indicator of otherwise healthy bodies, such selection would have paid back over time through the Darwinian algorithm. Well, unsurprisingly, evolutionary psychologists have tried to apply the idea to man, and one evolutionary psychology theory currently doing the rounds suggests that men are attracted to women with symmetrical breasts. "One of the few HBES members who still calls herself a sociobiologist, Sarah Blaffer Hrdy ... [*comments that many male investigators see*] the world through the filter of [*their*] own fantasies. 'Men just love coming up with scenarios for female breasts because they love looking at them,' Hrdy snaps", wrote John Horgan in *Scientific American*. This comment did make me chuckle, but possibly we may have found one area where there may still be a large Darwinian influence. You see, this is not the sort of thing that culture has ever sought to influence. Now if the Eleventh Commandment had stated

Magnificent exceptions?

"Thou shalt not lie with the woman of symmetry, because the shape of two perfect breasts is the mark of the Devil's whore", then *Baywatch* would probably have been a proscribed television programme hated by the 95% of Americans who believe in God. And women would be dressing from head to foot in heavy robes or getting silicon implants on just the one side. But Moses stopped at a round ten. And because he did – because, indeed, we are in non-proscribed territory – such hangovers from our Darwinian past may still exist.

The problem here is that I just don't know. Not because such ideas may not be plausible, but because the same ideas spring from the minds of evolutionary psychologists who try to put everything human down to adaptation. There are hugely important areas of study within evolutionary psychology. I just don't trust evolutionary psychologists to study them. To earn our trust in future many researchers are going to have to start being a lot less credulous than they have been in the past. There is huge pressure in the modern world for quick and simple answers. It comes from parents, politicians and pressure groups, as well as from scientists. But this is no excuse for poor quality thinking. Good quality thinking is vital, because we must have some form of trusted human Darwinism. Intelligent evolutionary psychology is crucial for policing behavioural genetics, if nothing else.

When a team of behavioural geneticists at the Institute of Psychiatry in London announced a couple of years ago that they may have isolated a genetic component to group behaviour, the work was worthy of serious consideration. Concentrating on the evolutionary psychology distinction between males and females, they claimed to have a genetic basis for why females often seem to be better at certain group work than males. From a Darwinian point of view perhaps there is indeed such a partial genetic component to human group organisation, because we do see different group behaviours in male and female chimpanzees. But to then use this to try to explain human group moralities, as some at the Institute of Psychiatry have done, is plain wrong. Close group working by female chimpanzees is just as much a product of their selfish genes as are the different forms of male behaviour. Female chimps may sometimes work in closer groups, or work longer in close groups, but they are no "nicer" than male chimps, no more decent, no more virtuous. Females will tear the arms off an unrelated chimp infant just as quickly as males will.

Magnificent exceptions?

So whatever the sins of their fathers, we still need behavioural Darwinians. Maybe by the time we get to the "son of the son of sociobiology", to extend Maynard Smith's terminology, we can start to relax a little.

Final thoughts
At the end of this chapter two important thoughts immediately spring to mind. They revolve around our similarities and differences.

Our differences
Firstly, our differences. Not all children diagnosed with ADHD have solely cultural origins to their problems. The great majority do, but some do not (and usually the cause is an inborn abnormality to the frontal lobes of the brain). The question then becomes: "So what?" Having a genetic susceptibility to hyperactivity is simply that; a genetic susceptibility to hyperactivity. It in no way predisposes one for a life of crime or the other nonsense that is often talked about. If there is a genetic tendency to hyperactivity, or risk-taking, or any other character trait, this has no moral equivalent. It is culture that will determine the different moral lives led. For example, if there is a genetic component to risk-taking it will be found as easily (and to the same extent) in soldiers, politicians, senior businessmen and stuntmen as it will be in criminals. All these groups have been identified as including numerous seekers of excitement and risk, and maybe some of this craving is genetically given. But whether one turns out a soldier, a businessman or a criminal will be up to culture, not whether one has the wrong allele of that "risk-taking gene". Biology may be a partial component to certain mannerisms; it is no component to morality. Morality is something that must be learned over many, many years by *all* of us. There is no morality coded into your genes, only gross immorality. Biology may make life harder for some children in specific cultures, but again, it is culture that determines who is going to "turn criminal".

A good example here for us to briefly turn to is dyslexia, an information processing disorder that most commonly manifests itself in difficulty in interpreting the written word, and also in memory problems. Dyslexia seems to have a strong genetic component[26], and

---

[26] Although as Robertson notes, quite how strong is "totally unclear". This is Medawar's

Magnificent exceptions?

there is some evidence that children with dyslexia are statistically at higher risk of turning to crime. But this in no way implies that a genetic predisposition to dyslexia is a genetic predisposition to crime, or that a dyslexia gene is a marker for a "criminality gene". Children with dyslexia can be just as smart as other children, but they often struggle in school because of their reading problems which need tackling in special ways. But unless their problem is identified, and relatively simple corrective actions are taken, school life can be very hard for such children. Such children may then act up in school to cover their own embarrassment and feelings of personal failure, so dyslexia can lead to behavioural problems. But the one characteristic in no way connects to the other. Let us hypothesise that a culture required not good written word ability, but good musical ability. That those with perfect pitch (which may have some genetic component) are valued, while those without such ability struggle through school life, often mocked by their more musical peers, perhaps even ridiculed by their teachers. Such children would feel embarrassment, and fear and loathing of school. Such children may learn to rebel. And I hazard that this would have been true of many of my well-educated readers; it would certainly have been true of myself (I have musical inability). In such cultures I would most certainly have found school life very difficult, and might very easily have rebelled against such constant personal denigration. Modern Western education is designed for those who are good with the written word, while other human cultures have required different abilities in their education systems. There have been human cultures that have praised artistry, or foreign language ability or good memory. Religiously fundamentalist countries tend to base their education around a specific holy book that

---

point that genetic influence is a function of nurture, and nurture a function of nature. Genetic influence in cases of dyslexia may only seem to be high *because* children received inadequate stimulation. Given more effective stimulation the genetic component may be far smaller. The impossibility of trying to give exact (or even ballpark) percentages is demonstrated by Steve Jones in *The Language of the Genes* when he discusses the medical disorder PKU. Here an inherited abnormality means sufferers cannot process a particular amino acid. Untreated such children have low intelligence and die young. But a simple change in diet to remove phenylalanine produces perfectly normal development. Try attaching genetic percentages in such a case - 100% or zero? "Its only answer is usually that there is no valid question", says Jones.

the child must then learn by heart. Today there are cultures where children are forced to learn and recite nothing but the lines of their scriptures hour after hour, day after day, year after year. The most cherished children are those who can perform such a feat of memory, the most disregarded and despised are those who cannot. Yet unlike the musical-based educational culture, I would probably have succeeded magnificently in such a memory-based educational culture because I have an excellent memory. Fortunately for this Darwinian, I did not grow up in such a culture, and just as fortunately I did not grow up in a culture that values musical ability. A gene for dyslexia is simply that: a gene for dyslexia. It is our cultures that determine what then happens to a child with such a learning disorder. Whether such children subsequently turn to crime depends on the culture, not on the individual. Morality has no genetic component. Morality is the result of our experiences in life, not our genes. Yet until we have the courage to see ourselves as products of the natural world we will continue to build our castles in the air, positing our genes for crime, for patriotism, for morality. And each time we posit genes that do not, *cannot*, exist, we fail those children who have real genetic disorders.

And just as importantly, even if there is a genetic component in some cases of ADHD, *who cares*? If some children can (from birth) only function most successfully with the help of a little extra teacher attention, or a tiny little pill, so what? We are all born cannibals. We all received just enough culture to overwrite our murderous biologies. We each required years and years of acculturation to make us what we are, equivalent to tens of thousands of cultural programming hours. So if young Sally requires 110% of this attention to get her to the same situation as most other children, or 100% plus a little pill, how dare you say there is something wrong with her? You got all the culture you needed to make you *you*. Any less and you wouldn't be you. A lot less and you would still be a cannibalistic ape. Luckily for you someone didn't draw the line a little earlier in your life, or you too would have found yourself the subject of moral panic. And luckily for you someone did not draw the line much earlier in you life, or you would be fit only for the Monkey House at London Zoo. Every human ever born has carried certain genetic disorders. They are called cannibalism and infanticide. If a few children have the genetic disorders of cannibalism, infanticide *and* attention-deficit, so what? Culture had to find solutions to the first two disorders, so why not to the third? What, you're genetically "normal" but they're not? I don't think so, Bonzo.

We all carry small biological differences, but it is our cultures that determine how any of those differences affect our lives and our opportunities.

Our similarities
My second thought is about our similarities. Major human behavioural differences are almost entirely cultural in origin because we share the same genetic nature. By now I hope you will have zero time for the core sociobiological belief that humans evolved virtue; that our ancestors broke the billion-year mould and evolved decency. However, your rejection of this evolutionary heresy does not entail you believing that all the major theories to come out of evolutionary psychology are necessarily totally wrong. Human virtue may be cultural, but this does not require that all human nastiness is necessarily cultural too. So, you might be thinking, perhaps Martin Daly and Margo Wilson are on to something when they claim that human infanticide has a genetic component, just as chimpanzee infanticide is genetically driven.

I don't think this idea is really plausible, for the reasons we have touched upon. To *accept* that culture is so powerful that 99% of the time it can tear us away from the grossly immoral behaviour coded into our genes, but then to *reject* the idea that cultural forces can manipulate us into nasty behaviour as easily as they manipulate us into nice behaviour, seems counterintuitive. To accept that culture can give us virtue, grief, and love, but not also vice, contempt and hatred is illogical. Much human infanticide has historically been ideologically inspired; those same myths which Darwin saw freeing us from our biology have been capable of promoting the murder of the children of a lesser God. So, as Dennett tells us, we have to be extremely careful when drawing parallels with natural world behaviour. Richard Owen, the nineteenth century "Establishment" paleontologist and *bête noire* of both Huxley and, in later periods, Darwin, detested evolutionary gradualism and (as with the evolutionary psychologists) saw the first man as having sprung fully formed from the womb of an ape ancestor. Like Daly and Margo Wilson today, Owen suggested that "original sin" might be the malingering ape in our make-up. Man good; ape bad. As with Owen, Daly and Wilson's belief that the good in us comes from our unique human "conscience

Magnificent exceptions?

module", while the bad in us can only come from our brutal common ancestor, seems simply another example of supreme human vanity.

A similar argument could be made for Randy Thornhill and Craig Palmer's hypothesis in their much-publicised *A Natural History of Rape*. Firstly, I must actually applaud them for (arguably) sticking to the evolutionary psychology "default assumption". By noting that natural selection dictates that *all* male primates are genetically programmed for rape they reason that in human males it is purely upbringing that can determine who will, or will not, rape. We share a common genetic inheritance, and so Thornhill and Palmer accept that it is culture that determines who will become a rapist: there but for the grace of a chance upbringing. Not for them the anti-evolutionary absurdity that rapists carry a "rape gene" while noble behavioural geneticists do not. (Although they cannot resist making the single behavioural genetics claim that a small subset of rapists - "psychopaths" - may be a genetically distinct "morph", or form, of human.)

But, secondly, I fundamentally disagree with their associated hypothesis that most human rape is therefore necessarily genetic in origin. Thornhill and Palmer refuse to see Huxley's paradox (as well as much sociological evidence) and again base their entire work on the heresy that conscience evolved in mankind. For them morality is still an adaptation, albeit an adaptation that must compete with other opposing adaptations, such as the rape instinct. Thornhill and Palmer are simply re-peddling Owen's "malingering brute" hypothesis: man good, ape bad. We will only finally kill this malingering brute conceit once we begin to accept that we are all born with the singular natural world genetic code for gross immorality, and not the genetic code for 1% natural world gross immorality and 99% anti-natural world human morality. Culture (acting on our very large brains) creates the wonder and the virtue. And it almost certainly creates most of the horror and the vice too.

But here's my point. Let me for a second be content with having won you half way over in rejecting virtue in nature. Let us therefore make a giant assumption and assume that Daly and Margo Wilson may be partly right, and we can *sometimes* equate non-ideological human infanticide with natural world "utility" infanticide. The question again is: "So what?" The *difference* between the behaviour of a step-parent who loves a stepchild and one who kills a stepchild is not genetic. We are apes, and we *all* carry the genetic code for infanticide (and, in men, for rape), and

Magnificent exceptions?

we will therefore all carry Daly and Wilson's hypothesised "malingering brute" in our make-up. It is not that one step-parent is a genetic freak, and one is not. It is that one was an ape who was lucky enough to have his or her infanticidal programming overwritten by cultural forces, and one was not. The second remained no more an ape; it still ended up reprogrammed for patriotism, religion, racism, et cetera, but just unfortunately not for love and tolerance. They were born identical. *Culture determined their behavioural differences.*

8

The "delusion" of Free Will

And thus we come to understand that the history of moral feelings is the history of an error, an error called "responsibility", which in turn rests on an error called "freedom of the will". ... No one is responsible for his deeds, no one for his nature; to judge is to be unjust. ... The tenet is as bright as sunlight, and yet everyone prefers to walk back into the shadow and untruth - for fear of the consequences.

Friedrich Nietzsche – Human, All Too Human

After spending a chapter in her book *The Ant and the Peacock* telling us what vassals humans are to our biology, Helena Cronin is careful to delineate the Darwinian kingdom. "Incidentally, all that we have just seen suggests that we shouldn't look on free will and biological 'constraints' as pulling in opposite directions. ... [I]ndeed, that far from constraining us, these 'constraints' are the very instruments of free will". In *The Red Queen* Matt Ridley is quick to support this understanding. "Besides, there is nothing inconsistent with free will or even chastity in this view of life", he writes. Yet these evolutionary psychologists are only following in a great tradition. "The paradox of determinism and free will appears not only resolvable in theory, it might even be reduced in status to an empirical problem in physics and biology", wrote E. O. Wilson in *On Human Nature* in 1978. It is a message Wilson has passed down to his followers. Critics of (human) sociobiology "mistakenly believe that sociobiologists are claiming all human behaviour to be so rigidly programmed by our genes that there is little or no scope left for free will", writes John Gribbin in *Being Human*.

There is only one slight problem. This is not Darwinism.

The general delusion about free will obvious. ... Believer in these views will pay great attention to Education.

Charles Darwin – Notebooks

Nineteenth-century materialism

Darwin was undecided about the existence of gods, but he was unequivocal in his rejection of free will. This is because Darwinism doesn't just tell us that culture, and not genetics, almost exclusively determines the important issues in human behaviour. It goes much further than that. It tells us that we are products of natural selection; it tells us that we are animals. It tells us that we must therefore obey the laws of physics, and not just the laws of nature.

In Desmond and Moore's epic biography *Darwin* the authors tell us that as late in his life as 1879 Darwin was to confess that he most certainly was not an atheist, and that he saw himself more and more as an agnostic. And so, at the grand old age of 70, Darwin was still undecided about *God*. But, you see, he was never undecided about *man*. *Gods* can exist outside the material realm of physical laws. *Man*, a product of evolution, cannot. So by 1838, when he was a mere stripling of twenty-nine - when it was to be another five years before he had a working understanding of his Earth-shattering theory of evolution by natural selection, and thirty-three years before he was ready to go into print on the human animal - he rejected the concept of free will. In his notebooks left behind after his death he was to write: "[O]ne well feels how many actions are not determined by what is called free will, but by strong invariable passions – when these passions, weak, opposed & complicated one calls them free will – the chance of mechanical phenomena".

The "delusion" of Free Will

Every action whatever is the effect of a motive. ... The general delusion about free will obvious – because man has power of action, & he can seldom analyse his motives ... he thinks they have none. ... One must view a wrecked man, like a sickly one ... - it would however be more proper to pity than to hate & be disgusted with them. ... This view should teach one profound humility, one deserves no credit for anything ... nor ought one to blame others.

Charles Darwin – (in Barrett et al., 1987, pp. 606, 608)

Yet his criticism of such anti-material philosophical doctrines was largely restricted to his personal jottings. Until the 1860s Darwin was terribly afraid that his new science would be seen to undermine the established Victorian social order. Even after 1860, ill health and a continuing need of social acceptability meant he was to keep many of his metaphysical conclusions to himself. But while the reclusive Darwin stayed publicly silent, he relished the combative stance of his disciples, and none more so than Thomas Huxley. Huxley, the great Darwinian orator and evolutionary propagandist, would lecture his huge audiences on the "automata" of nature. As the highest animal, man, Huxley thundered, was a conscious automaton, a thinking machine, moved by material causes unbroken by such mystical notions as "free will". Poignantly, Darwin's last words to his disciple of 30 years, penned in March 1882 just a couple of weeks before Darwin died, were: "I wish to God there were more automata in the world like you".

In the nineteenth century even Darwin's greatest critics were to realise that if man is truly a product of evolution by natural selection then free will is nonsense. On Saturday 30th June 1860 Bishop Samuel Wilberforce took on Darwin's Bulldog in the most celebrated debate in the Darwinian chronicles. This was the debate at the British Association for the Advancement of Science meeting in Oxford where, before a crowd of over 700, Wilberforce asked Huxley whether it was on his grandfather's or his grandmother's side that he was descended from an ape. Huxley purportedly responded that if he had to choose between having a miserable ape as a grandfather, and a man of considerable influence who uses his skills to the ridicule of science, he would unhesitatingly choose the ape. Yet "Soapy Sam", this theist lampooned and demonised by every Darwinian of the nineteenth century, was to demonstrate an insight that is today lost on many of Darwin's modern

The "delusion" of Free Will

sociobiological "followers". "[M]an's free-will and responsibility ... [*is*] utterly irreconcilable with the degrading notion of the brute origin of him who was created in the image of God", he wrote in his 1860 review of *On the Origin of Species*. The intellectual courage of Wilberforce's stance seems somewhat unusual by today's standards. Wilberforce did not like the implications of the theory of evolution by natural selection, and so he openly rejected the theory. He had a worldview that incorporated a number of things Darwinism denied; in his estimation the theory must therefore be wrong. But what he did not do was co-opt the theory, and twist it until it fitted his worldview.

Contemporary materialism

And if physical science, in strengthening our belief in the universality of causation and abolishing chance as an absurdity, leads to the conclusions of determinism, it does not more than follow the track of consistent and logical thinkers in philosophy and in theology, before it existed or was thought of.

Thomas Huxley – "Science and Morals"

The (philosophically) materialist understanding of Darwin is unchanged one hundred and sixty years later. "For a geneticist, all variance which is not genetic is, by definition, 'environmental'", writes John Maynard Smith. "[*Richard*] Dawkins, then, is a determinist, and so is every scientist who studies behaviour, even if they don't know it. ... What he would deny is the existence of something called 'free will' as an additional cause of behaviour" over and above genetics, upbringing, circumstance and contingency. In his essay "Popper's World", reprinted in *Games, Sex and Evolution*, Maynard Smith critically reviewed Karl Popper's *The Open Universe: An Argument for Indeterminism* which had at the time only recently been translated into English. "[*Popper*] points out that there is an apparent contradiction between the widely held view that every event has a cause, and the common-sense conviction that, at least sometimes, men are free to choose what they will do". The late Karl Popper had two great goals in his life. The first was to stop the abuse of scientific method and teach people how to differentiate between genuine knowledge and unfalsifiable pseudo-knowledge. The second was to provide a philosophical justification for free will. Popper attempted to achieve the latter by first showing that the simple predictable "clockwork" universe analogy is wrong – human systems become too complex to predict. "Today he would probably refer to the mathematics of 'chaos'", wrote Maynard Smith. But "chaos", as we shall see later, is deterministic. Unpredictability does not equate to indeterminism. And Popper accepted this. He, like the eighteenth-century philosopher Immanuel Kant, continued to acknowledge what every modern scientist accepts. This is that the material world inhabited by human beings is a world of prior causation. Therefore if you want to try to ground free will you must posit another realm. A *new* plane of existence beyond the purely material. Kant tried it with two worlds. Popper tried three.

"On the other hand it is equally necessary that everything that happens should be inexorably determined by natural laws", wrote Kant in *Foundations of the Metaphysics of Morals*. Free will was for Kant a paradox – the "antinomy" of freedom. Everything that exists in nature, in the empirical world, is bound by the laws of causal necessity - and yet we think of ourselves as *free*. To resolve this paradox Kant was to distinguish between the noumenal and the phenomenal, the former the "thing-in-itself", involving a timeless realm that lies outside the chain of causation, and the latter the knowable, empirical realm of causality and experience (things as they appear). This distinction causes great controversy regarding Kant's conception of the two realms. All we need to reflect on is that Kant, who as much as anyone wished to ground free will, could only even attempt to do so by going beyond the pure realm of the material. The same is true of Popper. Popper, in collaboration with John Eccles, proposed that we should recognize three "Worlds". World 1 is the phenomenal, the world of material things. World 2 is the world of consciousness. World 3 is the world of cultural artifacts, but in their conceptual rather than their material form. World 1 is "causally open" to World 2, and World 2 is causally open to World 3; but at some stage causality simply gives up the ghost. "The Popper-Eccles view, if I understand it (and if I do not, the fault is not entirely mine), is that at some point this chain of material causation is broken, and something immaterial causes a neurone to fire", Maynard Smith tells us. This Popper-Eccles idea would be vehemently denied by almost every modern neuroscientist. But, as Maynard Smith tells us with heavy irony even as he refutes their view, the Popper-Eccles view of neurones breaking free from causation is ultimately unfalsifiable: "[t]here is no way of showing that this is not so". From Popper's own methodological standpoint the Popper-Eccles hypothesis can only be viewed as pseudo-knowledge. For Popper and Eccles free will must become an article of faith, not science, a faith that simply denies the laws of causal necessity.

Unsurprisingly, given his hostility to unfalsifiable Just So Stories (Popperian or otherwise), Stephen Jay Gould had no more time for free will than his more gene-minded colleagues do. Echoing Maynard Smith's comment above he tells us that "Upbringing, culture, class, status, and all the intangibles that we call 'free will', determine how we restrict our behaviors from the wide spectrum ... that our genes permit" ("So Cleverly Kind an Animal", reprinted in *Ever Since Darwin*).

Darwinism forces us to face a very simple truth. We are products of the blind forces of gradual natural selection, we are material beings, and material beings live solely in the world of causal necessity (Kant's phenomenal realm). "Free will" is a delusion, as Darwin said, and so when the evolutionary heretics go into detail on their defences of free will, such as Edward Wilson in *On Human Nature*, it tends to be less than impressive. Wilson's "resolvable in theory" reduces to a misapprehension of quantum physics (specifically the Heisenberg uncertainty principle) used to buttress a misunderstanding of a basic mathematical distinction. First off, to deal briefly with the implications of quantum physics for free will. There are none. Free will theorists gave up trying to use quantum theory to support their beliefs years ago. The macroscopic universe we inhabit obeys deterministic laws, because of an effect known as decoherence that effectively reinstates classical mechanics at the level of systems above the very, very small. Far more importantly, and as has been noted by numerous philosophers, the random chance of quantum theory has no connection whatsoever to the concept of ethical freedom anyway; the freedom to *choose*, the freedom to *will*. Doing something because (hypothetically) a subatomic particle randomly moves inside your skull is no more "freedom" than doing something because genes or culture dictate it. The quantum event may be uncaused, but your (hypothetical) resulting action would itself be caused by the quantum event. The action is therefore not uncaused, and it is most certainly not chosen or willed. As the mathematician Norbert Wiener said in 1948: "The chance of the quantum-theoretician is not the ethical freedom of the Augustinian". (Do note that not all physicists are convinced quantum uncertainty truly predicts an indeterministic universe. In September 2002 Gerard 't

The "delusion" of Free Will

Hooft, 1999 Nobel laureate for physics, suggested that beneath quantum mechanics there might be a fully deterministic theory with information loss that "may also be essential for understanding causality at Planckian distance scales". Scientists might have been too quick to make quantum mechanics appear "to be mysterious", he suggests ('t Hooft [2002]).)

Wilson was using quantum uncertainty to support his main argument that human actions are unpredictable. And I have no problem with this basic argument. Human motivations are far too complex and interwoven to ever allow more than simple attempts at prediction. What is tragic, though, is that Wilson confuses *unpredictability* with *indeterminacy*. I can't accurately predict how a tossed coin will land because the variables are far too complex to track, but that does not mean that a coin toss is not a deterministic event. The variables, such as speed of flick, angle of flick, exact hand position at the flick, drag effects of the air, physical properties of the surface the coin will bounce on, etc., are too interwoven to attempt realistic prediction. But this does not mean the coin toss is not in theory predictable. Just that my little brain cannot do it. The relatively new science of "chaos theory" (also known under the working name of "deterministic chaos") deals with unpredictable, yet deterministic, systems. To quote the physicist Paul Davies: "The idea that a system can be both deterministic yet unpredictable is still rather a novelty". The reason, says Davies, can be traced to the system's extreme sensitivity to initial conditions. In the more familiar predictable systems of the sort studied in elementary mechanics, small errors in input description propagate to small errors in output. "In a chaotic system the errors grow exponentially with time, so that the slightest error in input soon leads to complete loss of predictive power" (in Davies [1989]). Chaotic systems, he tells us, have been found in an astonishingly wide range of systems, and some famous examples include turbulent fluids, weather systems, dripping taps, insect populations, electrical circuits and chemical reactions.

Finally, Wilson gives us a discourse about "feedback loops". Granted that most computer programmes (heck, even air-conditioning systems) involve feedback loops, Wilson's overall attempt at proving free will was remarkable for its inability to address the problem. I will leave the final word on Wilson's "proof" of free will to the philosopher Philip Kitcher, from his *Vaulting Ambition*. Most philosophers (including Kitcher) are as desperate as Edward Wilson in wanting free will, but this does not mean they will accept irrelevant pseudo-proofs. Hence Kitcher writes that Wilson's assault on the problem of human freedom simply bypasses the difficult questions. "The difficulties that arise from determinism, worries about historically remote determination or externally imposed desires, cannot be assuaged by misformulating them in terms of unpredictability and then taking consolation in the limited predictive powers of human beings."[27]

It is heartbreaking that twenty years after E.O. Wilson confused indeterminacy and unpredictability such a simple distinction is still being misunderstood. In his 1999 book *Genome: The Autobiography of a Species in 23 Chapters* the science journalist Matt Ridley concludes that: "Freedom lies in expressing your own determinism, not somebody else's". He appeared on Melvyn Bragg's *In Our Time* on BBC Radio 4 on 23 September to explain what he meant: "We have got free will because our determinisms interact; I mean my determinism interacts with yours and that ends up being thoroughly unpredictable". Rather frustratingly free will is still being conflated with unpredictability. And I was saddened that Bragg's other guest, the respected geneticist Steve Jones, supported Ridley's "clearly true" position. I have a lot of time for Jones as a technical geneticist, and in fact in November 1998 sent him a short letter apologising that I had ended my paper "Unnatural Selection" on a quote from him that, given the context, was a little unfortunate. Jones' view is that "we truly don't have the evidence" to be able to advance the behavioural genetics claims commonly made[28], and I wanted him to

---

[27] Kitcher himself seems to ground his free will in a belief that our evolved capacity for rational reflection somehow manages to break the causal chain. As for David Hume, free will lies in our *character*, and we should just ignore the bothersome detail that character is itself causally created. Most vertebrates reflect on their experiences and learn from past mistakes, as do complex computers. Does this therefore mean that sparrows and good chess computers also have "free will"?

[28] *The Observer*, 29 August 1999, page 18 (interview with Euan Ferguson). As Jones writes in *Almost Like a Whale*, "Th[e] mind is built from genes but what it can do has long transcended DNA". Yet this evolutionary fact doesn't stop those keen to give biological answers, and the infinite patterns of society "are justified with adaptive stories to fit. ... The new culture of the science pages [*of the Sunday papers*] uses a nodding acquaintance with evolution to promote an ethical agenda. ... It is the universal excuse".

know that I most certainly did not view him as a heretic. Both Ridley and Jones do share something in common, I believe, and this is a desperate *need* to believe in free will (although I think their probable motivations may be very different, coming as they do from opposite ends of the political spectrum).

Jones made a rather strange comment in *The Language of the Genes*. He noted the hugely vacuous (he prefers the terms infantile and foolish) "all-in-our-genes" worldview again in vogue, and the surprising reaction of the gay community to Hamer's "gay gene" claim. Jones is so busy fighting the public's extraordinarily naïve acceptance of untested genetic claims that it angers him when groups act in this way despite the lack of firm scientific evidence. "Nevertheless, the biologising of behaviour goes on apace. ... The next step - geneticising crime: 'It wasn't me that did it, it was my genes' - cannot be far behind." Jones tells us that all this is disconcerting to biologists, many of whom, he says, have spent years fighting the idea that crime, poverty or behaviour are inborn and cannot be altered by social means. "There is a grim Calvinism about the idea that talent, mental health or sexual preference are writ in DNA. But, for some members of the gay community, the idea seems to be more easy to accept than is that of unfettered choice." The strange comment was to suddenly add the last two words of the last sentence to the foregoing. Culture bequeaths us wonderful mental plasticity ("altered by social means"), but not choice. One is external to a system, the other an hypothesised internal piece of magic. Jones, I believe, is so desperate to clean genetics of the "all-in-our-genes" abuse that he has allowed himself to mix two distinct problems (a mistake that unfortunately reappears in his *In the Blood*). Jones appears to believe that cultural programming somehow "sets us free", which is something it most definitely does not do. While it frees the human race from the dictates of genetics, it does not free the individual from the dictates of (genes and) culture. Programming is programming, whatever its source. Whether human behaviour is "all-in-our-genes" or whether this view utterly disregards the huge role of culture *has no bearing at all on the free will debate*. The free will debate is not a debate about the relative power of genes over culture, or culture over genes. Free will is a totally separate issue. The free will debate is a debate about whether we have a *third* option available to us free from both genetic programming and cultural conditioning (or any complex mix of the two). New genetic

The "delusion" of Free Will

knowledge actually adds nothing whatsoever to the millennia-old free will debate. Genetics is just a contemporary way to say "nature". The nature/nurture debate has been going since before the time of the philosophers of ancient Greece, and it is quite different from the equally ancient free will debate. Modern philosophers of science understand this distinction (see almost anything by Dennett or Philip Kitcher). Darwin, too, understood this.

Given the extraordinary spectacle of materialist Darwinians celebrating free will, even the evolutionary heretics are now starting to change their tune. There has recently emerged a new breed of evolutionary psychologist who has had to admit that "free will" is not a scientific concept. "[I]t means 'not caused by anything,' and the scientific worldview can only seek causes", said the evolutionary psychologist Steven Pinker in an interview for the popular magazine *Wired* in March 1998. Pinker has realised that even under the misguided Darwinism of evolutionary psychology the concept of free will becomes extraordinary. Consequently in his book *How the Mind Works* he rejected the concept and has proposed that ethics and knowledge should be seen as separate spheres: free will does not exist, but we must *pretend* it does. "Like many philosophers, I believe that science and ethics are two self-contained systems played out among the same entities in the world, just as poker and bridge are different games played with the same fifty-two-card deck", he writes. The science "game" treats us as material objects, while the ethics "game" blesses us with an imaginary "free will". Pinker's argument, in slightly different forms, has been used by *status quo* apologists for millennia. It is the view of a privileged, white Western male; one feels that someone born (through no choice of their own) slightly less fortunate might see things a tad differently.

Pinker has effectively announced to the world that the blame-based and exceedingly punitive American legal system cannot stand the scrutiny of science. Parochial justice, like faith, must not be subjected to the light of reason; we ought to forget philosophical duty and continue to live in the cave and forever see (or profess to see) only shadows[29]. If there was ever a more pressing reason for the fundamental re-

---

[29] In *The Republic* Plato provided us with his simile of the cave to demonstrate the ascent of the mind from ignorance to knowledge. Most people were like prisoners chained in a

examination of a particular "morally based" legal code, I don't know what it can be. At least Plato's prisoners had the excuse that they neither knew better nor had any help to leave the cave. With all due respect to certain professors at MIT, Plato's prisoners did not simply decide to be troglodytes.

Smuggling free will in by the back door

But it is not only evolutionary psychologists who want free will. Even that otherwise relatively orthodox Darwinian philosopher Daniel Dennett is desperate to keep free will in the picture. Interestingly, he attempts to do so by the semantic trick of changing its definition.

The main theme of *Elbow Room: The Varieties of Free Will Worth Wanting*, Dennett tells us, is to demonstrate that philosophers have been wrong for thousands of years to argue that choice is necessary for a person to have free will. In the midst of all the discord and disagreement among philosophers about free will, there are a few calm islands of near unanimity, he explains. "As van Inwagen notes: 'Almost all philosophers agree that a necessary condition for holding an agent responsible for an act is believing that the agent *could have* refrained from performing that act'. ... But in fact, I will argue, it is seldom that we even *seem* to care whether or not a person could have done otherwise." This "could have done otherwise" principle is "simply wrong", he tells us (see also Dennett [2003]). Dennett's argument seems to be that philosophers are wrong to think choice important to the free will debate, because for thousands of years uninformed mystics have believed otherwise. Philosophers shouldn't rock the boat and should just accept the tabloid reader view of life. Dennett further admits that it is not only his readers who may desperately want free will. He tells us that he also so wants to believe in free will, and "what one hopes very much to be true may be true". As a philosophical work this is totally unacceptable - philosophy is supposed to be about truth; it is not about

---

darkened cave since childhood, who knew nothing of the world beyond the cave except for the illusory shadows of the outside world thrown on the cave wall. Yet even were the prisoners to be set free they would be so dazzled by the light outside that most would retreat to the comfort of their dark cave. So the philosopher is the former prisoner who has been forced to walk forward and stare into the light, seeing things for how they really are.

tabloid opinion, and it is most certainly not about hopes and wants.

In *Darwin's Dangerous Idea* he (briefly) tries again to separate choice from free will by redefining "freedom". The Viking spacecraft, we are once more told, had free will as soon as radio messages took so long to reach it that it was cut loose from NASA control. Free will is again defined purely in terms of mechanical self-control, and not freedom to have chosen differently. So long as an entity has self-control it has free will. Free will is therefore possessed by apes, mice, ants, mould, bacteria and viruses. Under Dennett's formulation were you to take your son's toy car, put in new batteries and then set it to race away, it would not have free will. However, as soon as you turn your back and walk away from it never to return, you have blessed it with free will. So, according to Dennett, even the Energizer Bunny can have free will. I can't help feeling that Kant would not have been overly impressed with Professor Dennett and his Duracell definition of freedom.

Dennett is actually giving another slight variant on the highly contentious philosophical doctrine known as compatibilism (pioneered by the eighteenth-century philosopher David Hume, and sometimes called "soft determinism"). Compatibilism purportedly *accepts* that people's decisions are wholly determined, but argues that determinism and free action are compatible, without ever attempting to square this circle. Kant called compatibilism a "wretched subterfuge". It is a pre-Darwinian doctrine that simply disregards the point that people, like all machines, cannot choose to do differently. It tries to smuggle free will in through the back door by simply ignoring uncomfortable material fact.

Yet nervous Darwinians are not the only ones today to try to smuggle free will in through the back door. Even orthodox philosophers can play this game at the drop of a hat. One example is the political philosopher Susan Wolf who wrote a paper called "Sanity and the Metaphysics of Responsibility" (which you can find reprinted in Christman's *The Inner Citadel*). The paper began by conceding defeat in the "caused-or-not" debate; Wolf thereafter tried to change the rules. Causality is admitted, but this is no longer what free will is taken to mean. Free will is now all about "sanity". The fact remains, though, that (once causality is admitted, as with Wolf) "sanity", like unpredictability, has nothing whatsoever to do with freedom of choice, with freedom to will, with "free will". Wolf is hiding from the implications of modern biology in dogma that dates from the ancient world. Insanity has historically

The "delusion" of Free Will

tended to be viewed as a hindrance to free will; insanity (it was argued) limits free choice, so sanity presumably evinces free will. The original flaw, of course, is in the first statement: insanity is not a hindrance to free choice as free choice (as Wolf admits) never existed. You cannot keep an ancient tenet ("the insane have reduced free will, while the sane have unimpaired free will") when you have abandoned the worldview upon which it was built (belief in the possibility of uncaused actions, and the defining of free will in terms of free choice).

Philosophers like Dennett and Wolf will continue to tie themselves up in metaphysical knots for so long as they try to defend the indefensible. "Free will" is a spurious belief ceded us by our mystical and unthinkingly deferential past, and evolution will no longer be co-opted to defend a particular political line. *Pace* Professor Dennett, free will is about freedom of choice, and free will is, as Mr. Darwin said, a delusion. So what are the implications?

"And therefore, my dear Thrasymachus, justice is the interest of the lucky"

The message is clear: those who will not accommodate, who will not temper, who insist on keeping only the purest and wildest strain of their heritage alive, we will be obliged, reluctantly, to cage or disarm, and we will do our best to disable the [beliefs] they fight for. ... Child abuse is beyond the pale. Discrimination is beyond the pale.

Daniel Dennett – Darwin's Dangerous Idea

"What is justice?" famously asked Plato in the best known of his dialogues, *The Republic*. The Greek word here translated as "justice", *dikaiosune*, also meant doing right, or moral action. "I say that justice or right is simply what is in the interest of the stronger party", stated the character Thrasymachus, giving one of the definitions of justice common among ancient Greeks. He meant that justice or morality is simply that code of behaviour imposed on everyone else by those with power, and that there is no worthier explanation to be invoked. Plato's Socrates immediately countered that, while in barbaric states administered justice may be the interest of the strong, in well run states with fine laws and traditions administered justice would be synonymous with the ideal of justice. "No science studies or enforces the interest of the controlling or stronger party, but rather that of the weaker party

The "delusion" of Free Will

subjected to it. ... And therefore, my dear Thrasymachus, no ruler of any kind, *qua* ruler, exercises his authority, whatever its sphere, with his own interest in view, but that of the subject of his skill". Proving that justice is something both noble and achievable has always been of overriding concern to philosophers, from Plato through to Kant and beyond. After all, justice is what sets us apart from the animals, is it not? Yet Thrasymachus was wrong. Justice is not the interest of the strong. That at least would have a certain *nobility* for some philosophers. Nietzsche, for example, would have applauded such justice. But the truth, I fear, would be humiliating even for Nietzsche. Justice is not the interest of the strong. In modern states with laws and traditions, justice becomes the interest of the *lucky* [30].

While gods may live outside the laws of physics, man, a product of evolution, may not. Like all other animals, we are biochemical machines. We human beings are the most complex machines that exist in the known universe, the product of two often fundamentally opposing programming languages. One programme is set at conception and the other is added to, deleted from, rewritten and overwritten every second we live. Two hugely complicated programmes that then combine with unimaginably complex (but deterministic) interaction. This scientific understanding does not make love any less wondrous, or grief any less painful, or life any less precious. A sunset becomes no less beautiful to the human mind by knowing that it is simply radiated energy in the observable spectrum moving through an atmosphere; a rainbow is no less fabulous just because we can speak of refraction (although, importantly, it does give us no right to expect the pot of gold). Accepting that free will is a delusion simply means that we are machines that can begin to grow up.

Why do people behave the way they do? Simply because that is how they have been *programmed*. We are built as apes, but run with billions of different cultural programmes. Apes are simply indifferent; no grief, no real love, no hate. Cultural reprogramming is subsequently responsible for both the kindness and the monstrosity in the human world. But genes and culture - that's really all there is. And justice

---

[30] Certain sections of this chapter were first used in a paper I wrote for *Children & Society*, the journal of the National Children's Bureau. Copyright © 2000 John Wiley & Sons Ltd.

becomes the interest of those more lucky in their programming.

Yes, abandoning our adherence to free will is going to have profound repercussions[31]. But this understanding does not mean, for example, that we cannot take damaged children and put them where they cannot harm others while we try to repair that damage. We need to segregate such damaged children until they can become useful and non-threatening to our societies. We can tell them that they have done terrible things. We can tell them that their current behaviour is unacceptable in our societies. *But we cannot blame them.* They exist because we did not care enough to raise them in the image we wanted. They exist because we looked away for the decade or so it took to build them. They had *absolutely no choice* in how they would turn out, and no way to prevent their future actions. While we could have ensured that a person turned out differently, he or she could not. No one can change himself from the *inside*. Change can only come from the *outside*. We are the products of the programmes we have at any one time, and if you want to change that current behaviour you need to change the programmes. To take out one set of instructions and substitute another. Genes cannot easily be modified. This, however, makes little difference. Human moral behaviour now has almost nothing to do with our genes. Civilisation is, as Huxley realised, an artificial construct made possible by the cultural overwriting of our biology. Cultural programmes can usually be modified, except where the prior programming has gone too deep to modify. Sometimes changing those programmes will be too much trouble, or is beyond our limited ken. But

---

[31] It is important to point out that the implications of a rejection of the idea of free will are *not*, as Dennett puts it, "almost too grim to contemplate" (*Elbow Room*). In "Science and Morals", Huxley described as "that absurdity" the idea that the belief "in the uncausedness of volition is essential to morality". Our entire upbringing makes us what we are, not simply an attachment to a particular piece of tacky mysticism. Additionally, rejecting metaphysical freedom by no means impairs social freedom. We have the power to write the future. But it is a power that is forever *constrained* and that is ceded to us only by dint of knowledge about ourselves and our past. Knowing what we are changes what we are. By *accepting* our limitations we are actually freer than if we cling to mumbo-jumbo about "free will". Darwinism can give us a world of less violence, less hypocrisy and less unfairness. But we will get there by playing by the scientific rules, not by tearing up the rulebook.

blaming the person because, when we finally make the effort to begin to undo years of bad programming we find that it is then too difficult, seems to me to be little short of contemptible and absurd.

Children in England are only held to be responsible if they can be shown to "know" the difference between right and wrong. But such knowledge neither does nor does not determine behaviour. Such knowledge is only a single "sub-programme" within any child's immensely complex and often hugely inconsistent programming. The solemnly intoned question of whether the child knew what it was doing was wrong is therefore a largely fatuous line of inquiry that does far more to demonstrate legislators' misconceptions about the physical world. Again, some us are lucky enough to have been majority programmed under good circumstances, while the unlucky get majority programmed under poorer circumstances. So for the lucky to feel *superiority* over and *anger* towards the unlucky? Well, to me this is discrimination that I cannot condone. Feel anger at the act, feel a burning need to try and change the cultural programmes that create the unlucky losers in our system (*our* victims, and, consequently, *their* victims), but how dare you feel superior to these people, or somehow better? You have been *lucky*, that is all. There but for the grace of a more hardworking programmer.

The horribly ironic thing is that our children who turn criminal have not actually failed us. They have no choice but to follow their cultural conditioning. They therefore do not fail us, but we fail in our responsibility to help them change. Violence and criminality are not the signs of an especially sick individual (only an unlucky individual), but they are the signs of a sick society. It is up to us to begin to accept this, and begin to undo the harm we have done. We can tell such children they have done very great wrongs, but we cannot blame them. All we can really say is "sorry". In parts of Africa they say it takes a village to raise a child. Well, it also takes a village to abuse a child[32].

Morality as a Process

Free will is a delusion. But is it a necessary delusion? As T. S. Eliot said, "humankind cannot bear very much reality". Should we therefore

---

[32] Oh, and if you are the sort of person to be swayed by either Dennett or Pinker's reasoning (or indeed by any form of compatibilism), please note one thing. Neither

separate morality and truth, as Pinker wants? As a Millennium gimmick, the influential science agent John Brockman got together over one hundred leading scientists and science writers and asked them to nominate what they saw as the greatest invention of the last two thousand years. John Horgan chose free will: "Science has made it increasingly clear ... that free will is an illusion. But - even more so than God - it is a glorious, absolutely necessary illusion," he wrote.

This view of the world is called consequentialism. Consequentialism argues that the merit of an action must be determined by its consequences, not by the character from which it arises. The most famous philosophical doctrine often criticised as a form of consequentialism is utilitarianism, the ethical system elaborated by Jeremy Bentham in the late eighteenth century. Utilitarianism founds all practical reasoning in the concept of utility, or maximising the happiness of the community. Utilitarianism is attractive to many in political philosophy because it seems to offer the hope of a truly coherent and systematic moral system; a system that does not owe allegiance to any particular preconceived belief or depend on some contingent point of view. The weakness of utilitarianism is that, in their understandable craving for an objective table of values, philosophers have been attracted to a system that is capable of excusing startling injustices. As the political theorist Will Kymlicka puts it: "utilitarianism could justify sacrificing the weak and unpopular members of the community for the benefit of the majority". Horgan and Dennett are both giving consequentialist justifications for clinging to the belief in free will. And the real problem with Dennett's consequentialism is its ethical dimension. Or rather, its lack of an ethical dimension.

Like gods, free will has been at times a useful tool in the social armoury. But, like gods, it can leave terrible injustices in its wake. Under Dennett's special pleading, the substance of the free will debate is totally lost. Dennett's apologia deliberately avoids addressing the issue, critical to the question of *fairness*, of whether or not a person could have

---

Dennett nor Pinker's arguments make *one jot of difference* to the fact that, under any evolutionary worldview, justice is the interest of the lucky. A choice-less causal chain is a choice-less causal chain, however you dress it up (Dennett, Wolf, Hume, etc.) or ask us to ignore it (Pinker). By rejecting the idea of blame Darwin had the courage - intellectual and moral - to face up to what so few philosophers have the courage to face up to.

chosen to behave differently. If people can't choose to do otherwise then there can be no blame, just as Nietzsche said. It becomes unfair to blame them for something they have no control over. It's like blaming a crippled child for being crippled.

Now there are still people in the Western world who blame crippled children for being crippled, who say that these children's afflictions are a punishment from God. But there are far fewer of these people now than once there were, thanks largely to Christianity. A thousand years ago many Western societies blamed the mentally retarded for being retarded. Villagers said the mentally ill were being punished by God for their past sins. It was the Christian Church that put a stop to this most hideous mindset. It was the Christian Church that educated people to the opposite conclusion; that the sick and weak should be protected, not vilified. The philosopher Daniel Robinson, who has spent much time researching the history of feelings toward the mentally ill, says that it was only with the arrival of "great kings and good clerics" that the attitude to the mentally disturbed began to change.

Of course abandoning our belief in free will does not mean that we cannot imprison people while we try to change their behaviour. Darwin did not believe in free will, but he still believed that society needed to be protected from the people who would do it harm. But the flip side to that is that there are many people - and especially children - in our society today who need to be protected from the mendacity that says they have chosen a life of crime, they have chosen a life of begging or sleeping rough. Until our legal systems accept that free will is a delusion *our laws and our court processes will continue to be founded on lies and discrimination.*

We live in a world where politicians and judges and newspaper proprietors and even philosophers get to propagate unfairness and injustice, while the moral organisations of the world, such as the Churches, currently look the other way. But, strangely enough, while I have no faith that politicians or judges or newspaper proprietors or most philosophers will change their ways, I do have confidence that theologians will. Consequentialism has little hold over the religious mind, because clerics, unlike philosophers, do not care about utility, and want only to know about the issue of fairness. The doctrine of free will is Plato's Noble Lie writ with the blood of a million children, and you cannot begin to build God's kingdom on earth until you have fairness at the core of your religion.

Unlike philosophers, clerics cannot afford to dismiss science, because clerics ignore science at both their own, and humanity's, peril. Religion is today valuable because it offers the opportunity for human moral advancement. The day religion proves itself incapable of this is the day we no longer need it. Christianity is about protecting those who cannot protect themselves, and since we are machines, since all change can only come from the outside, this means that clinging to a belief in free will in the face of the scientific evidence is a *profoundly* anti-Christian act. These children, the weak and despised children of the world, need the Church's protection far more than the children who have been luckier in their programming. It seems that Darwin was actually a far better Christian than the Church has given him credit for. We can agree with Bishop Wilberforce (1860) when he says: "Mr. Darwin writes as a Christian, and we doubt not that he is one". So the question has never been: Is Mr. Darwin a Christian? The question has always been: Are most Christians Christian?

"... With liberty and justice for some"

Incidentally, I thoroughly agree that free will is a stupid idea.

G. C. Williams – (personal communication)

Yes, Darwinism often makes uncomfortable reading. Yes, Darwinism often challenges received opinion. This is why Darwin wrote to his disciple Joseph Hooker that it was "like confessing a murder"; this was why Darwin fêted Huxley as "my good and kind agent for the propagation of the Gospel - *i.e.* the devil's gospel". Huxley was the one Darwinian who never backed away from a fight in his life and, while Darwin was cautious of offending the great and the good of Victorian Britain, he applauded Huxley's uncompromising stance.

Darwinism makes you take a long, hard look at the world around you. And at the justice system of regions that seem to produce evolutionary heretics by the dozen. In the American state of California, the "three strikes and you're out" rule operates. Californians are supposedly given three chances, and those who break the law for a third time (involving even the most minor felony, including the theft of just a few dollars) are automatically locked away for the rest of their lives. California was once in the vanguard of American liberalism. During the 1980s and 1990s, however, prisoner rehabilitation spending was slashed in the Californian penal system. California, like many U.S. states, willingly bought into the myth of behavioural genetics; the idea that criminals are born not made. The idea that Californians get any "chances" is anti-Darwinian nonsense. Criminals are *made* in California, not born (of course, even if they were born this way they would still have no chances as free will does not exist whether culture or genetics holds the upper hand). So until an outside force makes a serious effort to unmake these unlucky losers, *nothing* can change for these people. Truth and honour are supposed to be important to Americans, but there is neither truth nor honour in the vicious hypocrisy of pretending people have opportunities they do not, while slashing all attempts to help such people. Science teaches us that truth, honour and common decency do not exist in the Californian legislature and judiciary, so with dozens of U.S. states having introduced similar legislation it is perhaps unsurprising that American philosophers and politicians wish to dismiss evolutionary knowledge, or even ban it altogether.

Maybe you can begin to see why I not only detest the anti-Darwinism of the evolutionary heretics, but I quickly lose patience with free will apologists like Daniel Dennett and those who appear to be satisfied with Darwinism *lite*. Dennett may not "even [] care whether or not a person could have done otherwise", but like philosophers throughout history Darwin did care, because Darwin understood the philosophical implications. Justice, if it is to have any claim to objective value and decency, must be something more than just the interest of the lucky, the interest of those more fortunate in their programming. If justice serves only the interest of the lucky then justice *dies* as a philosophical ideal. The ideal of justice has traditionally been associated with the philosophical ideals of fairness and objective morality, and in a post-Enlightenment world the law is supposed to be built upon such concepts. As the philosopher Daniel Robinson put it in his *Wild Beasts & Idle Humours*, a survey of three thousand years of legal accountability, "[l]ess controversial in this regard would appear to be the implicit logic of every system of law: that the reach of law extends only to creatures able to ... abide by its prescriptions". For as long as our systems of justice represent *criminals'* victims but refuse to represent also *our* victims (the criminals themselves) justice is not impartial. For so long as justice remains the interest of the lucky, the interest of only those programmed *to be able to abide* by the law's prescriptions, justice dies as a philosophical ideal. No legal system represents all citizens. And no judge is ever fair. Nietzsche concurred: "No one is responsible for his deeds, no one for his nature; to judge is to be unjust". Darwin even went so far as to draw a parallel between the way humans treat criminals and the ignominy of nature. In his notebooks he wrote: "Animals do attack the weak & sickly as we do the wicked - we ought to pity & assist & educate".

When Dan Robinson seems to argue, with Dennett, Wolf and Pinker, that whether freedom of choice even exists is *irrelevant* to the law he unconsciously demonstrates the poverty of the argument. No philosopher of law who holds to this position has ever been willing to put the argument to the test by campaigning to remove this admitted fiction from the law, or even by campaigning to ensure the public is told of the fiction. My own view is that only a recognisably medieval legal system (of the type that used to blame the mentally and physically handicapped for their handicaps) could ever *openly* admit people have no choice yet still get away with treating them with vicious brutality and contempt. Hence it becomes of overriding importance for certain legal systems to maintain their fictions, even as their philosophers of law maintain theirs. It is, of course, precisely because Darwinian materialism does not suffer fools or hypocrites that Darwinism is a threat to so many people. In their biography of Darwin, Adrian Desmond and James Moore recount how an 18-year-old Darwin was present when William Browne, a President of the Plinian Society at Edinburgh University, argued that mind and consciousness emerge from matter, from the brain. Browne's argument caused outrage amongst conservatives who saw mind as a spiritual gift. The poet Samuel Taylor Coleridge saw the speech as "subversive" and the *Quarterly Review* demanded that it be legally suppressed. Yet today Browne's understanding is at the heart of consciousness theory and all modern science. Perhaps the Californian legislature will now also attempt to "legally suppress" Darwinism? The truth is, after all, so awkward to any system that hides behind child abuse, hypocrisy and deceit.

It is tragic enough that we raise so many children in damaging conditions. But to then *blame* our creations when they act a certain way because this is how *we have built them*? What arrogance it is to despise what one has created. To wash our hands of our mistakes and pretend it

The "delusion" of Free Will

was nothing to do with us, all the while hiding from the truth in bad science and cheap mysticism. We take children incapable of stopping us, the most vulnerable children who have absolutely no one to fight for them, and we force them to become criminals. Then, like all bullies, like all criminals, like all child abusers, we make up self-serving rationales to try to excuse our atrocious behaviour toward our victims. Dennett's admission that he finds a universe without free will "almost too grim to contemplate" just illustrates the problem here. Discrimination does not stop being discrimination because those in positions of influence would prefer not to reflect on the truth[33]. "Free will" is the last great Western prejudice, and justice, my dear Thrasymachus, is the interest of the lucky.

Justice isn't blind. According to Darwin she's just stupid and bigoted.

This view should teach one profound humility, one deserves no credit for anything ... nor ought one to blame others.

Charles Darwin – 1838

ENDNOTE TO CHAPTER 8

Damned Man Walking:
Christianity Meets Darwin's Wager
Writing around 1660, the French mathematician and philosopher Blaise Pascal gave theology one of its greatest arguments in favour of belief in

---

[33] An example of the rare philosopher who is capable of thinking like a scientist is the American scholar Derk Pereboom. In *Living Without Free Will*, Pereboom echoes Darwin when he concludes that "many of us have a strong and visceral feeling that it is unfair to regard causally determined agents as deserving of blame. ... More than this, [*the truth*] demands a certain level of care and attention to the well-being of criminals that would radically change our current practice". Pereboom ends his book on a striking quotation of Albert Einstein's that reminds us of why we need scientists just as we sometimes still need philosophers: "I do not at all believe in human freedom in the philosophical sense. Everybody acts not only under external compulsion but also in accordance with inner necessity. ... This realization ... prevents us from taking ourselves and other people all too seriously; it is conducive to a view of life which, in particular, gives humor its due" (Einstein [1954]).

The "delusion" of Free Will

God. Pascal was the first Western thinker to replace the traditional epistemic approach to belief with a purely pragmatic one. Pascal's Wager goes as follows. If reason can tell us nothing about the existence of God (the position of the agnostic), and yet there is even the remotest chance that God exists, then it always pays to wager in favour of God. If you wager for God, you stand to gain an eternity of happiness if God exists, while you risk losing nothing if He does not. If, however, you wager against God, you stand to gain nothing if God does not exist, while you risk losing everything if He does. Wagering on God is always the best bet.

Let us weigh the gain and the loss involved by wagering that God exists. ... If you win, you win all; if you lose, you lose nothing. Wager then, without hesitation, that He does exist.

Blaise Pascal – Pensées

	God Exists	God Does Not Exist
Wager that God is:	Gain everything	Lose nothing
Wager that God is not:	Lose everything	Gain nothing

Pascal argued that logic forces the wager on us precisely *because* there is uncertainty over the existence of God. His special contribution was to realise that the wager works equally well whatever probability you assign to God's existence, provided that the probability is non-zero (and non-infinitesimal). The mathematics always returns an expected utility to belief in God of infinite, because whatever probability you assign to the existence of God (above 0 or infinitesimal) multiplied by the expected payoff (gain everything, or "infinite") the return is always infinity. In contrast, a wager against God can only ever return an expectation of *minus* infinity, because whatever probability you assign to God (above 0 or infinitesimal) multiplied by the expected payoff (damnation if you are wrong) the return is always minus infinity. Rationality requires you to wager for God every time.

The "delusion" of Free Will

$$E \text{ (wager for God)} = (p \times \infty) + ((1 - p) \times f) = \infty$$
$$E \text{ (wager against God)} = (p \times -\infty) + ((1 - p) \times f) = -\infty$$

Where:
E is expected return,
p is the probability of God, and
f is any finite number.

Pascal's Wager can of course be criticised on many fronts, and has been down the centuries. In the mid eighteenth century Diderot pointed out that an imam could reason just as well this way, while Voltaire suggested that the Christian God would never be swayed by such calculating self-interest. But given that Pascal's Wager continues to exert such a strong hold on many Christian minds, it might be interesting to see what the new scientific knowledge brings to the Wager.

Darwin's Wager

Science cannot disprove the existence of Gods but it can disprove the existence of what we call "Free Will", and hence while Darwin never rejected belief in God he rejected without hesitation belief in free will. And this therefore raises a rather fascinating insight: *Pascal's own reasoning now forces us to extend his Wager*, yet in extending the Wager we also manage to avoid most of the criticisms that can be levelled against the original Wager. For this reason, we shall call our more rigorous theological argument *Darwin's Wager*.

> *Yes; but a bet must be laid. There is no option: you have joined the game. Which will you choose, then? Since a choice has to be made, let us see which is of least moment to you. ... But here there is an infinity of infinitely happy life to win, one chance of winning against a finite number of chances of losing, and what you stake is finite.*
>
> **Pascal** – Pensées

Darwin's Wager uses exactly the same logic to argue against a belief in "free will" *for anyone who also believes in God*. Our contribution to theology would be to realise that Darwin's Wager works equally well whatever probability you assign to free will's existence, provided that the probability is not 1 (absolute certainty). The mathematics always

The "delusion" of Free Will

returns an expected utility for belief in free will of *minus* infinity, because whatever probability you assign to free will (below 1 / absolute certainty) multiplied by the expected payoff (damnation for those who blame children for the decisions they are *forced* to make) always returns the expected utility of minus infinity (damnation). In contrast, a wager against free will can only return an infinite expected utility, because whatever probability you assign to free will (below 1) multiplied by the expected payoff if you are right (an eternity in heaven) always returns the expected utility of infinity.

Darwin's Wager (the If God Exists Variant) shown mathematically:

E (wager for Free Will) = (p x f) + ((1 - p) x -∞) = -∞
E (wager against Free Will) = (p x f) + ((1 - p) x ∞) = ∞

Where p is now the probability of free will.

So long as there is even a *tiny* chance that free will does not exist, then, if you are a Christian, rationality *requires* you to wager against "Free Will"!

	Free Will Exists	Free Will Does Not Exist
Wager that Free Will is:	Gain nothing	Lose everything
Wager that Free Will is not:	Lose nothing	Gain everything

As already mentioned, Darwin's Wager also manages to avoid the main objections commonly levelled against Pascal's Wager:

The Many Gods Objection. As Diderot explained to us, Pascal's Wager can tell you nothing about which God to believe in. However, this is no objection to Darwin's Wager, as all Gods are presumed to be interested in truth.

The Self-Interest Objection. Voltaire's objection was that at least the

The "delusion" of Free Will

Christian God would never be swayed by such indecent and puerile pragmatism. Again, this is less of an objection to Darwin's Wager, because whereas the simple profession of belief (wagering for God) may say nothing about an agent's underlying moral worth, by wagering against free will the agent actually goes a long way towards evincing his or her moral worth, and commitment to truth.

The Truth Problem. One problem with Pascal's Wager is that it holds that you lose little or nothing by wagering for God, if God does not actually exist. In fact the costs of losing the truth could be large, ranging from a blind parroting of dogma, to injustice, or even wasting precious time on this planet. Darwin's Wager avoids these objections, because it takes no view on the truth of something that cannot be falsified (God), but deals only in the truth of something that can be falsified (free will).

The Reasonable Assumptions Objection. Pascal's Wager works only under the highly debatable assumption that the probability of God's existence is greater than 0. Darwin's Wager, in contrast, requires only the far less contentious assumption that the probability of free will is less than 1.

All the evidence is that "Free Will" does not exist. Free Will is therefore a matter of faith, not reason. Have faith in God, said Pascal, and you do not risk an eternity of torment. Have faith in Free Will, says Darwin, and you'd better be prepared for eternal damnation.

Let us weigh the gain and the loss involved by wagering that Free Will exists.
... If you win, you win nothing; if you lose, you lose all. Wager then, without hesitation, that Free Will does not exist.

Darwin's Wager

9

Lighting up hell-fires in Christendom

…that the foundation of morality is to have done, once and for all, with lying; to give up pretending to believe that for which there is no evidence.

Thomas Huxley – "Science and Morals"

The first time I ever saw Richard Dawkins lecture, he had offended a theist within ten minutes of starting to speak. Dawkins, besides being a noted evolutionary theorist, is a renowned militant atheist. Okay, it was at a British Humanist Association Centenary Lecture (at the London School of Economics on 20th June 1996), and the lecture was entitled "Science as Religious Education", so I sort of expected some fireworks. That crowd was out for religious blood (if you thought the religiously inclined had a monopoly on intolerance you would be mistaken) and Dawkins did not disappoint. I can't even remember which particular religiously-excused horror Dawkins was relating, but it was enough of a slight for the religious heckler to storm out in protest. Unsurprisingly, given the crowd, he was not followed out of the hall.

Now I fear religious fundamentalism as much as Dawkins. I look around the world at the horror done or excused in the name of religion and I feel physically sick. But this chapter is about more than showing that ideologies can so easily lead to monstrosity. This chapter is about the contingency of those ideologies, and it is about how a Darwinian can never have a simple reaction to religion or other ideologies. A Darwinian will always feel ambivalence toward religion, because he or she understands that those same religions that can have such a pull on the human mind that they become so easily twisted into horrible forms actually *made* man in the first place. These are the stories that saved us from the horror of our genes. But by making us human, they make us human in all its forms. By saving us they also damn us. This is the contradiction we must address if we are ever to begin to move forward.

Darwinism has often being characterised as another factor in the slide into moral relativism. That the world is moving from a system of universal values buttressed by unchanging religious law to one where there is no established order; where morals have no universal validity and are valid only in relation to particular social circumstances, particular societies, or particular individuals. The truth could not be more different. While Darwinism will always bring difficult ethical challenges with it, based as it is on knowledge and not blind faith, it also gives us the hope of the first truly universal moral code man has ever known. Darwinism can offer us a world stripped bare of the subjectivism, relativism and hypocrisy so often found within established religion, and can give us instead access to the only objective moral code that exists. Darwinism can peel away not just the hypocrisy of anti-evolutionary beliefs like "free will", but also the hypocrisy of cultural relativism. Darwinism actually offers us both hope for the future, and our first chance to live by truth and not falsehood.

Far from subsuming the social sciences within biology as the sociobiologists have sought to do, biology carves out a separate sphere for the social sciences, in the same way that biology is carved from the sphere of physics. Culture allows us to escape our biology. Biology can tell us everything we need to know about what we are escaping from, but it can tell us little about what we escape to: the contingent world of the social scientist. For ten thousand years we have been the products of our myths and, because of this, we have been capable of extraordinary brutality. We have split the world into those unholy in the sight of gods, or unholy in the sight of the gene, and we have killed the reprobate. We

Lighting up hell-fires in Christendom

have created relativist ethical systems based in religion, or politics, or nationalism, or pseudo-science. The only ethical system we have *not* tried is the one based in the unchanging truth about ourselves.

All the wonder

Most of us, if we see somebody in great distress, weeping - we will go and put an arm around them and try to console them. It's a thing that I have an overwhelming impulse to do ... and so we know that we can rise above our Darwinian past.

Richard Dawkins – (in interview)

We do not need to spend much time considering the wonders that cultural manipulation has been capable of because this has been the running theme of this book. We know the effects; the taking of just another cannibalistic ape living in groups of 60 and the creation of the first human. Our susceptibility to cultural manipulation took an ape living in extreme viciousness, in small untrusting groups, and allowed it to be bound into groups of many thousands, groups of many millions. It took an ape incapable of much in the way of co-operation and enabled startling levels of co-operation, co-operation many have described as "ant-like" in its scale. Culture took a diploid creature and produced haplodiploid behaviour. It allowed us to create all that you see around you. It built cities and spawned civilisations. It built empires and sent men to the moon.

But it produced so much more than purely mechanical co-operation. Cultural manipulation made possible all that is best in our species, all those behaviours that have no place in a Darwinian world. It created grief, love, honour and duty. It allowed us to care and be cared for. It allowed us to love and be loved. It allowed us, too, to be worthy of love. It allowed us to continually strive for something better, a chance to forever improve ourselves. Culture saved us from mere existence and gave us lives worth living.

But the wonder hides a terrible truth. The terrible truth about our susceptibility to the dark side of cultural manipulation as well as to the not so dark. Darwinism explains why we are so susceptible to the dark side of culture in the first place; why our salvation is also our damnation. In *The Selfish Gene* Dawkins wrote in an oft-quoted passage: "We have the power to defy the selfish genes of our birth and, if necessary, the selfish memes of our indoctrination. ... We, alone on earth, can rebel against the tyranny of the selfish replicators". In the parlance of Richard Dawkins "memes" have transcended genes, and Dawkins believes that by training our children in reason and empathy we can help them to escape the blind dogmas of religion, nationalism and political ideology that have in their turn allowed us to escape our genes. "We" don't have to repeat the mistakes of our forefathers. But "[t]his 'we' that transcends not only its genetic creators but also its mimetic creators is, we have just seen, a myth", writes Dennett in *Darwin's Dangerous Idea*. Dawkins is not trying to tell us that "we" have a third inheritance system, "free will", that gets to dictate to both genes and culture. Only that we should not be pessimistic because both genetic effects and, more importantly, cultural effects can be changed. Yet the ontological problem then becomes *what restricts* these powerful gene-defying ideologies so as to inculcate only goodness? What comes along to change the cultural programmes already set if those programmes happen to be evil? What exists, separate from the ideologies, to stop them doing harm? And the answer is: Nothing. Only more culture can change culture. This, said Darwin, is why good education is so vitally important. Because we have no third option - nothing that gets to stand "outside" genes and culture - we are at the mercy of the gene-defying cultural programmes we have been deeded. And often the cultural cure has been worse than the genetic illness. At least when chimpanzees slaughter, they don't moralise about it at the same time.

And all the horror

... I for my part will undertake to prove that rape, murder & arson are positively enjoined in Exodus.
Thomas Huxley – (in Desmond's Huxley, p. 253)

In his novel *1984* George Orwell defined a concept he called "doublethink". "Doublethink means the power of holding two contradictory beliefs in one's mind simultaneously, and accepting both of them", he wrote. Doublethink was a mental characteristic of the population found in his totalitarian society. Philosophers have often drawn attention to the extraordinary behaviour of whole populations; Nietzsche, in *Beyond Good and Evil*, wrote that: "Madness is something rare in individuals – but in groups, parties, peoples, ages it is the rule".

Lighting up hell-fires in Christendom

Nietzsche realised that our myths make us what we are. That if we were to suddenly see the myths as myths it almost certainly doesn't change our core values, because by that stage our personalities are very often too deeply set to change. This understanding is something that so annoyed Nietzsche. He realised that humanism is little more than secularised Christianity, and since he venerated pre-Christian "master-morality" - the nobility of the barbarian, the proud spirit and indifference to others' suffering of the Viking – he wanted an end to Christian values. But the myths make us, mind and spirit, and when the myths are, or become, twisted, we not only become twisted, we cannot see the madness of our own actions. We cannot see the horror, or the mind-numbing hypocrisy.

Cultural manipulation is of course not solely benign; it may allow extraordinary altruism and kindness, but it also allows barbarity and horror. Cultural manipulation sets us free from genetic control, but leaves us at the mercy of cultural control, "victims to our memes" as Dawkins has so often put it. Victims to cultural manipulation. Indifferent killing for gene survival is the way of the natural world. Enthusiastic killing for gods, countries and ideologies is the way of only the human world:

And the children of Israel took all the women of Midian captives, and their little ones. ... And Moses was wroth with the officers of the host. ... And Moses said unto them, Have ye saved all the women alive? ... Now therefore kill every male among the little ones, and kill every woman that hath known man by lying with him. But all the women children, that have not known a man by lying with him, keep alive for yourselves. And do ye abide without the camp seven days. ...[P]urify both yourselves and your captives on the third day, and on the seventh day. ... And the booty, being the rest of the prey which the men of war had caught, was six hundred thousand and seventy thousand and five thousand sheep ... and thirty and two thousand persons in all, of women that had not known man by lying with him.
The Holy Bible – Numbers 31, verses 9 to 35.

Most Darwinians do not spend a lot of time reading *The Old Testament*, so some may not have come across this particular passage before. Those who have will have just been reintroduced to the horror. It becomes a little difficult to deny what's going on here, although, dear God, have

Lighting up hell-fires in Christendom

people tried. Let us consider the scale and the nature of what is being discussed. Now to the best of my knowledge in a pre-gynaecological age there was only one way to know for certain if a girl child had lain with a man. Please understand, it did not involve asking her. When you are doing this for your God you cannot afford to have one of these little pagan bitches lie to you. So Moses' men would have probably "inquired" of every young girl between the ages of thirteen and, let's say, ten (this was the pre-modern age, after all). We are also talking about slavery. "Keep alive for yourselves." Economic slavery at best, sexual slavery at worst. The point about slaves is they have to earn their keep. So girl babies (whose mothers, remember, are soon going to be dead) are just as much dead meat as boy babies. They would be just another mouth to feed, and utterly incapable of helping around a man's home. So, my best guess using the figures as given. 32,000 girls left alive, and I'm assuming the top age is twelve, and the bottom age six. Translates to perhaps another 35,000 girls under six exterminated, to add to 70,000 or so boys between the age of 1 day and twelve years (over twelve they probably died with their fathers in the earlier battles). Plus, assuming a very high (natural) mortality beyond forty and high child mortality rates, 100,000 women (at least 15,000 of whom would have been pregnant) and girls between the ages of twelve and sixty-odd. Over 200,000 boys, girls and women butchered, and 32,000 girls enslaved. And these figures include the slaughter of over 20,000 children under two, and the "inquiring" of maybe 15,000 young girls aged between ten and thirteen.

This was not an isolated incident. According to the Bible Moses ordered the annihilation of at least eight nations. The Midianites were only one of eight. In addition there were seven others bragged about in *The Old Testament*:

> *But in the cities of these peoples that the Lord your God gives you for an inheritance, you shall save alive nothing that breathes, but you shall utterly destroy them, the Hittites and the Amorites, the Canaanites and the Perizzites, the Hivites and the Jebusites, as the Lord your God has commanded; that they may not teach you to do according to all their abominable practices which they have done in the service of their gods.*
> **The Holy Bible** – Deuteronomy 20, verses 16 to 18.

In addition to the Midianites, the Hittites, the Amorites, the Canaanites, the Perizzites, the Hivites and the Jebusites, Moses commanded the extermination of the Girgashites (Deuteronomy 7, verse 1). Oh, and for any of you who tend to feel more sympathy for animals than you do for your fellow human beings, you should note that Joshua, who finally completed Moses' genocidal instructions, left his own particular mark by hamstringing all his enemies' horses (Joshua 11, verse 9). Can you honestly question why Nietzsche called us insane?

Yet Moses is not just a Jewish prophet[34]. He is a great shared prophet of many of the world's religions, and a few secular ideologies too. He is the greatest lawgiver and ethical prophet in human history, invoked and venerated daily by over half the world's population. He is also possibly our greatest serial genocide. The last two sentences taken together tell you everything you really need to know about the effects of cultural manipulation on the human ape. Throughout our long, sordid history, most people have not thought it always wrong to rape children or smash open the skulls of infants. Only to rape *certain* children, or smash open the skulls of *certain* infants. Midianite children and infants don't count for most people. The effect of our contingent worldviews tends to be to define so narrowly who is part of one's "group", and to declare how inconsequential are the sufferings of those outside the group.

Newman's intuitive spirituality made [Darwin] uneasy. ... How could faith in "the Holiness of God" have arisen amid the death, famine, and wars of Semitic tribes? No, Charles insisted, the religious instinct had evolved with society. The primitive Jewish God, whose atrocities had "lit up hell-fires in Christendom", could be nothing but a barbaric tyrant.

Adrian Desmond & James Moore – Darwin, p. 377

The strings that can never be cut

If we are ever to protect our children from ourselves we need to begin to understand the contingent insanity that cultural manipulation can introduce. Throughout human history there have been leaders who have cared for their people, and sought what was best for them, but

---

[34] Many secular scholars doubt that Moses ever existed. Does this weaken my case? Probably not. That half the world parrots the genocidal instructions of a prophet is bad enough. That half the world may be parroting the genocidal instructions of a fictitious prophet is perhaps even worse, if you think about it.

have found their answers in barbarity well beyond anything found in the animal kingdom. A three thousand-year-old model of pious moralising, followed by abominations almost beyond belief. Yet we have found it impossible to recognise such hypocrisy because all too often we have bought into the hypocrisy and the doublethink. It becomes a part of what we are; just another sub-routine deeded to us by our cultural conditioning.

I want to point out that, **because there is nothing beyond the two programmes of genes and culture**, we will be incapable of seeing the insanity until the 10,000-year-old cultural programmes begin to change. This is an understanding of almost unimaginable importance, yet it is an understanding that we run from in terror. There is no "us", separate from our selfish genes and our bigoted cultures, to do the seeing. And none of us are totally free from potential insanity; none of us get to step outside our programming. Insanity. That we can grieve for the Jews, and yet blindly ignore, or even justify, the suffering of the Midianites. To understand, at last, why no one weeps for the Midianites. I want you to begin to realise why mankind produces so many small-minded hypocrites, of all creeds and nations and races, who will talk of the need for "justice" or "morality" while applauding the torture and extermination of other people's children.

I do not seek to offend, but my concern must be for our future, because until we understand ourselves we cannot change. Moses' infanticide was not natural world infanticide. This was *human* infanticide, ideological infanticide to the greater glory of a God, "memic" infanticide if you wish. This, too, is cultural manipulation. Nature works through blind, pitiless indifference to the fate of other gene vehicles. Yet humans are almost never indifferent to the fate of others, neither when we are being incredibly kind nor when we are being incredibly cruel. I'm sure Moses' heart was singing as he did God's work. But as our forefathers escaped their Darwinian past by cultural reprogramming, we can use education to try to allow our children to escape the contingent insanity inherent within that reprogramming. They may then be able to achieve a degree of sanity probably never before witnessed in our species.

In knowledge, salvation?

Darwinian knowledge can make a difference for the good, although we must be aware of the dangers inherent in that knowledge. The Darwinian understanding is not only immensely powerful for understanding why we are the way we are, it changes what we are. Maybe it will help save just one nation out of eight from the horrors of cultural manipulation. If you have the strength to begin to see the parallels between a Hitler and a Moses, then maybe, just maybe, you will be less willing to excuse the next slaughter your party, or religion, or country wishes[35]. Christianity died locked inside the skull of Jesus Christ. Judaism largely died inside the skull of Moses. The legacies live on, and the legacies do not have to repeat the mistakes of the past. We get to write the legacies for our own time. For good or ill, the reprogramming is up to us, knowing what we know of the past. Darwinism teaches us about the reprogramming, and in so doing *changes* the inheritance.

I don't deny the potential for damage, and that the Darwinian worldview could have other maybe less positive consequences. The philosopher Nietzsche was the first thinker to confront the confusion that follows when a world, post-Darwin, is stripped of the old religious and authoritarian certainties. Nietzsche sought solace in Schopenhauerian escapism before turning to his programme of self-mastery and freedom of the spirit. Whatever we might think of his construction of a new system of values to replace "lost" theistic certainties, Nietzsche correctly understood the potential threat (both social and psychological) when old certainties collapse. Darwinian knowledge could ultimately weaken the social unity that was so important to the ancient Greeks, and which Socrates was put to death for threatening. Loyalty to the *polis* was everything for the Greeks, and nothing could be allowed to undermine it – especially not truth, as Socrates was to find out. The loyalty engendered by the old Greek myths made the Greeks capable of utter savagery to non-Greeks, but it did allow them to operate

---

[35] No behavioural geneticist has ever tried to suggest that Moses (or any other genocidal Old Testament prophet) was a biological freak, or that he carried the "gene" for serial killing, rape, arson, child murder, theft, etc. Any common criminal is genetically flawed but not this conservative hero who committed crimes many thousand times worse and is lauded by (presumably identically programmed??) conservative groups. The support for such behaviour surely indicates "a severe and dangerous personality disorder", but John Turner does not ruminate upon the indefinite imprisonment of such apparently dangerous people.

in relatively stable, cohesive societies. One person who considered this point from an evolutionary perspective is the psychologist (the late) Donald Campbell. Campbell was elected President of the American Psychological Association in 1975 and in his inaugural address he decided to deal with a pet interest of his. He saw important implications for psychology arising from evolutionary biology. His views were dismissed by some psychologists simply because they were new ideas in a profession fearful of change. And others saw him as just another sociobiologist. But Campbell was a very special sociobiologist, because he was in the tradition of Darwin, Wallace and Huxley. Unlike E. O. Wilson, he treated man as (genetically) just another ape. "I see urban humankind as the only vertebrate that approaches the social insects in self-sacrificial altruism." Because in humans there is genetic competition among the co-operators "this extreme sociality cannot have been achieved on a genetic basis. For vertebrates that share humankind's genetic predicament, the degree of sociality and altruism achieved by some baboons ... may well represent the maximum social co-ordination and altruism achievable", Campbell noted. (His inaugural address was published in *American Psychologist* - see Campbell [1975]).

By a process of elimination, Campbell, like Darwin and Huxley, saw culture as having to combat biology. Nevertheless, while Richard Dawkins sometimes seems to make the mistake of seeing only the negative side to cultural manipulation, Campbell makes the opposite mistake and sees only the positive and comes close to arguing that this makes tradition a good in itself, irrespective of its effects on both its adherents and other groups. Just because we would be cannibalistic apes without cultural manipulation does not mean that the cure cannot sometimes be judged to be worse than the original illness. As I have tried to demonstrate, a Darwinian must take a far more ambivalent attitude toward tradition and religion than either Dawkins or Campbell seem to take. Culture must be assessed on its own merits. Nevertheless, Campbell makes an important point about the cohesive value of myths and traditions, and the dangers of simply sweeping them away. Darwinian knowledge will always give legislators pause for thought. When politicians talk of the need for "education, education, education" they usually mean that we that we must stop before the child realises that the Emperor has no clothes. Our children must not be better informed than our leaders, after all. And many governments will always want to know that when they call on their peoples to exterminate Midianite babies they will have volunteers galore. So this may be another paradox: unity may best be achieved at the expense of both knowledge and sanity. Maybe by shining the brightest of lights into the cave, Darwinism will continue the "rot" that so many conservatives claim the Enlightenment brought to the world.

Because we can now see ourselves simply as the (very special) apes that we are, we no longer have the right to expect that the cosmos is in perfect balance; that we can have truth *and* still expect that this is the best of all possible worlds. "There is no pre-established harmony between the furthering of truth and the good of mankind", wrote Nietzsche in *Human, All Too Human*. Darwin, too, realised the potential for changed attitudes if we can but see that there is no such thing as free will. Some people, because of their weak prior programming, and changed by the amended programming that the new view brings with it, could potentially see the lack of free will as a new license to do harm. Yet Darwin's (rather Aristotelian) conclusion was that "[t]his view will not do harm, because no one can be really *fully* convinced of its truth except man who has thought very much, & he will know his happiness lays in doing good & being perfect". But of course, this "virtuous" man of Darwin's is himself as much a product of his cultural programming as is the non-virtuous, unless we are to believe that reason somehow sets us free to be virtuous. Aristotle thought so, but Aristotle is perhaps history's most notorious apologist for slavery and was indifferent to the suffering of non-Greeks. I'd not call this particularly virtuous, nor much of an advertisement for "reason".

Where does this leave us, then? What alternatives do we have if we cannot have truth and strong cohesion simultaneously? Should we live forever in ignorance, forever a hair's breadth away from genocide and other ideological horrors? Or should Darwinism be a secret knowledge made available only to the few, to the "intellectuals" and to those who govern us? Darwinism shows us the contradiction between the need for order and the intellectual groundlessness of much of that order. Tradition has little objectivity beyond the contingency that created it.

Darwinism is not prescriptive. But Darwinism is a tool to gain access to the truth about ourselves. And this truth itself is objective. Maybe the only objectivity man can ever truly achieve is the only form of objectivity that he has never tried in the 10,000 years of settled human existence. It is not the "objectivity" of myth. It is not the "objectivity" of tradition. It is the objectivity of self-knowledge, the objectivity of truth about ourselves. To go beyond - not good and evil as Nietzsche wanted - but hypocrisy. To stop quoting "Thou shalt not kill" even as we butcher Midianite babies. I believe that Darwin's legacy, when truly accepted, while it may face setbacks, cannot in the long run help but create fewer adherents to any fundamentalist ideology that so often seems to end in tears.

"for good or ill ..."
I think it was the French philosopher Henri Bergson who said: "The history of mankind is a quest for civilisation, a move away from barbarism". Once slaughtering the children of the unholy was just, and once slavery was just. Then our attitudes changed. We have tried to base our contingent ethical beliefs on just about every conceivable myth; myths about divine intervention, myths about national superiority, and myths about man's nature. The only foundation we have never tried is an attempt to base an ethical system in truth. The unchanging truth about the contingent nature of man. The unchanging truth that if we wish an objective morality we must begin to strip away the ideology. Darwinism is no threat to religion or tradition, only to fundamentalist religion, and fundamentalist tradition. A Darwinian, more than any other, recognises the crucial role of cohesive beliefs. But Darwinism will forever be a threat to religion or tradition that preaches hate and contempt and extermination.

A myth is a contingent programme, created unconsciously at one period to satisfy the cravings of that period. In both that originating "phenotype", and a later "phenotype", it can have consequences not previously anticipated. It can make us kind, and it can make us monsters, often at the same time and in the same person. As Dennett told us, we don't get any choice in the matter, there is no transcendent "me" that gets to choose separate from my existing genetic and cultural programmes. This understanding cannot be overstressed. I can't choose to take only the nice bits of my culture, and reject the rest. I need to have been lucky enough for someone or something to have set me up this way to so want to decide. I began life as a selfish ape. Along the way people and ideologies told me who and what to care for and who and what not to care for. So yes, there will always be contingency. But the first step to minimising our contingent insanity – to stop being victims to our memes as Dawkins puts it - is to admit that we have a problem. Darwinism forces us to face the problem, and in doing so can make us better people. It is difficult to be a Darwinian - a *true* Darwinian - and not a humanist, I believe. When you realise there is no free will, that we all begin life as selfish apes, that we are moulded and manipulated into our various human forms, and that this is our common inheritance, you will be either an absolute misanthrope or a humanist. Suddenly nationalism, political and religious ideologies and all other forms of artificial boundaries between people seem, not unimportant, but less important, because the certainty that others find in these contingent worldviews can be met through knowledge. And maybe, after ten thousand years, this is the best we can hope for.

I have a hope that one day we will have grown up enough to build holocaust memorials, not just to the Jews, but to all the victims of ideology and genocide. Including the Midianites. Cultural manipulation allowed us to transcend nature and create incredible wonders. It allowed us to build cities and fly to the moon. But there was a high price to be paid by so many. A price to be paid by us all. Civilisation and barbarous insanity: two sides of the same coin. The paradox of man.

The capacity of our species, for good or ill, to be swayed by myths I find a continuing and as yet unanswered puzzle.

John Maynard Smith – Games, Sex and Evolution

Man is not the altruistic ape. Man is just another selfish ape. But, crucially, man is also the contingent ape. For good or ill, we are the contingent ape.

10

"Take me to your Prophet"; and other essays

I can calculate the motions of the heavenly bodies, but not the madness of people.

Isaac Newton

"Take me to your Prophet"

Man has always been fascinated by the possibility of intelligent life on other planets, and by the great question of whether we are alone in the vastness that surrounds us. We have tried to imagine what, if they exist, these other creatures might be like. And yet all the time the answer has been staring us in the face. Modern Darwinism is set to answer this question that has so haunted us, and the answer will not be pleasant.

Our opinions about intelligent extraterrestrial life have always varied with our human concerns. On the one hand we have pined for the alien saviour. The American astronomer Percival Lowell captivated the popular imagination at the turn of the nineteenth century with his theories about intelligent life on Mars. As Carl Sagan wrote in *Cosmos*: "Lowell's Martians were benign and hopeful, even a little godlike". Yet on the other hand we have feared the alien threat. Ruthlessly indifferent to our very existence, with "intellects vast and cool and unsympathetic", as Wells wrote of his alien menace in *The War of the Worlds*.

Both views are wrong. Other would-be spacefarers will be neither rational and benign, nor rational and malign. They, like us, will be irrational mystics and ideologues, capable of awe-inspiring kindness one minute, and murderous zealotry the next. Man was not made in the image of God. But all intelligent life will be made in the image of man.

Darwinism and the selfish extraterrestrial

But is there anything that must be true of all life, wherever it is found, and whatever the basis of its chemistry? ... Obviously I do not know but, if I had to bet, I would put money on one fundamental principle. This is the law that all life evolves by the differential survival of replicating entities.

Richard Dawkins – The Selfish Gene[36]

Richard Dawkins calls it "universal Darwinism"; the hypothesis that only Darwinian principles are capable of explaining the organised complexity of life. And yet if Dawkins is right that life anywhere will obey Darwinian rules of evolution, then this means we have to face up to something rather important.

If life can only exist on the basis of competing (i.e. selfish) replicating entities such as genes, then this would seem to have the implication that John Maynard Smith's evolutionarily stable strategies, or ESSs, must hold true everywhere in the universe. These "strategies" are the possible group behaviours nature has discovered within the limits set by given genetic inheritance systems. For example, natural selection has ensured that close co-operation is only possible in small groups – for all other apes groups of around sixty to one hundred – and you cannot build

---

[36] See also chapter 11 of *The Blind Watchmaker* or Dawkins' 1983 paper "Universal Darwinism" which assess and reject the various suggested alternatives to Darwinian inheritance in the development of life.

civilisation with small groups. To build civilisation something has to overcome that (first) selfish replicator, and in man it is susceptibility to ideological manipulation that has mainly allowed us to overcome our selfish genes. The implication?

If life exists elsewhere in the universe, and *if* Dawkins is right about the universal truth of Darwinism, then that life cannot in the first instance be particularly intelligent (no real language; nothing beyond daily survival). The reason for this is that such creatures - answerable solely to their equivalent of DNA - will only be able to co-exist closely in small groups (assuming they have inheritance and reproduction models fairly similar to those found in Earth vertebrates). Small groups cannot evolve civilisation. They cannot evolve detailed language, complex culture, or spaceflight.

We are predominantly free of our genetic programme. We can therefore evolve detailed language, complex culture and spaceflight. The problem is that the same factors that let us transcend our genes leave us at the mercy of cultural reprogramming. Darwinism explains how and why we have ideologies and genocide. Unfortunately it also tells us that any other species that has developed language, complex culture and spaceflight will be just like us. Oh, not physically, of course; just mentally. They may have four tentacles instead of two arms. Yet they will use those four tentacles to hold their bibles, to pass out their political tracts, and to make their instruments of torture.

Reading their history books will be like reading our own. Like us, they too will have internecine wars, religions, nationalism and political ideologies. They, too, will have tortured and killed millions of their fellow beings for being unholy in the sight of their gods, or for heretically claiming that a prophet wore the wrong colour underwear. Every form of killing we have devised will be known to them, and all the while they shall excuse their vicious savagery by invoking gods, nations and political ideologies. Like us, they will have interrogated anatomy, physiology, psychology and physics to devise ever more imaginative forms of torture and slaughter; they, too, will have raised pain, mutilation and cant to an artform. And all the time, of course, with each act of small-minded savagery claimed as being in accordance with the principles of "justice" and "right".

So why do we need to travel to the stars? We are probably already there.

Conclusion
The last twenty-five years of human evolutionary theorising are largely wasted years. The implications of selfish gene Darwinism are immense, had we only the courage to apply the understanding to ourselves. Is the above true? As diploidy nature creates groups too small to allow for complex interaction until something begins to transcend biology, must all extraterrestrial intelligences share our madness? Haplodiploidy nature seems to offer the hope of large-scale close interaction without the need to transcend biology, until one realises that we must still conjure with the gross immorality of genetically given behaviour. As Dawkins writes: "True warfare in which large rival armies fight to the death is known only in man and in social insects".

Contingent insanity or blind, pitiless indifference. Is this the rule to all life anywhere? Probably, though I cannot be certain. But at the very least, like Wallace, I am asking the right questions.

The thinking reed

Fragile reed as he may be, man, as Pascal says, is a thinking reed: there lies within him a fund of energy, operating intelligently and so far akin to that which pervades the universe, that it is competent to influence and modify the cosmic process.

Thomas Huxley – Evolution and Ethics

The question under consideration here is whether reason can somehow take us into a new realm where we are no longer at the mercy of the contingencies of culture. This question always raises heated exchanges. There is one group of social scientists that tries to argue that humans are already sufficiently rational, and that most of our choices are already governed by calculated reason. Such claimants try to convince using computer simulated game theory models (such as Iterated Prisoner's Dilemma) that produce the ultimate rational choice in a given social interaction. While such models are sometimes useful, I often feel that the more exaggerated claims of game theorists are just another attempt to claim a special place for our species. Another frightened wish or conceited belief that we are especially blessed of nature. But the debate is very old, and often very bitter, so let us consider it in more detail.

Huxley himself believed that ultimately our (evolved) rationality - operating over progressive cultures - had been enough to show us that our evolved natures must be combated, that nature's moral indifference must be countered. The main problem with the argument may be that reason is a tool of evolution and all mammals possess a degree of it. Mice can reason their way around a maze, and chimpanzees use reason to work out how to block the escape routes of the mother of the infant they are preparing to kill. And yet we feel absolute horror at such natural world behaviour as we encountered in earlier chapters. The understanding that this is our genetic code does not just unsettle us, it will have sent many people screaming into the arms of the evolutionary psychologists. But why should this be the case if reason is just another tool of natural selection? As Peter Singer notes (without supporting) in *The Expanding Circle*: "[r]eason may help us to sort out the consequences of our choices, but it cannot tell us what we most want. 'Reason is, and ought only to be,' wrote Hume, 'the slave of the passions'". Hume's statement, argues Singer, is recognisably the ancestor of the view of human behaviour taken by some sociobiologists. "'Reason is, and ought only to be,' they would rewrite Hume, 'the slave of our genes.' On this basis they would dismiss as philosophical fantasy the idea that reason draws us toward a universal point of view." Philosophers have at least stopped spelling reason with a capital "R", but I am still inclined to think that the idea of a universally altruistic point of view accessible through reason comes close to being another example of human arrogance. As Stephen Jay Gould has put it: "Before Freud, we imagined ourselves as rational creatures (surely one of the least modest statements in intellectual history)". Darwinians have fought long and hard against the age-old claim that nature was one Great Chain of Being, with lowly insects at the bottom, and man almost at the top, just below angels and avatars. This was the old religious view, and it falls before the scientific knowledge that all creatures were selected to occupy particular evolutionary niches. The idea that the human animal was selected with the capacity for near perfect reason seems to me to be another crass attempt to separate ourselves from the natural world.

Of course, for many since the Enlightenment it may have been mentally a very attractive idea - reason becomes our saviour now that gods can no longer be. But our "wants" are surely set by our programming - genes and culture (and predominantly culture or we'd all keep looking for "infant" on the menu). Our genetic programming is indifferent to the gross immorality of nature, and our cultural programming is contingent. Learn to cry buckets at the death of Jewish babies, but happily use Midianite babies as firelighters. So why should reason automatically make us wish to follow a moral code that is the direct antithesis of nature?

Singer tried to make the argument that as culture made dealing with an ever more diverse range of groups necessary, ethical systems had to expand to encompass these new groups ("the expanding circle" of consideration for others). But this very nicely skirts around the problem of how we moved from being in selfish ape-like bands of sixty into being in the position of *wanting* to trade with others on relatively friendly terms; how we evolved true ethical systems in the first place. I've got some respect for Singer, but this just doesn't wash for me. I think he's reasoning *backwards* from where we are now – the altruistic human - to how he thinks we must have been to get here – the altruistic ape, perhaps? What we've got to be able to do is reason *forward* from the selfish ape to the altruistic human, and this is going to mean dealing with Huxley's Paradox first.

There is still a lot of thinking to be done on the role for reason. George Williams gave Singer's idea lukewarm support in his 1988 paper before giving his own main answer of manipulation through language. Dawkins too occasionally seems to be arguing for some role for an autonomous reason – "[w]hy are you and I so much nicer than our selfish genes ever programmed us to be? ... Brains as big as ours can actively rebel against the dictates of the naturally selected genes that built them", in Dawkins (1996) where he briefly discusses the role of rational foresight. For both Williams and Dawkins - and indeed Darwin - reason may be playing a part, but only a small part, in making us what we are. The reason that it cannot be playing a large part is that not only do we still have a lot of explaining to do as regards our visceral attitudes towards natural world behaviour – to our own genetic code - but because trying to find evidence that humans actually often employ their foresight or reason is rather a stretch.

John Maynard Smith pioneered evolutionary (game-theoretic) modelling through mathematical simulation, and he sees no problem with large–scale human cohesion if behaviour remains rational. As the eighteenth-century philosopher Kant wrote in his essay "Perpetual

Peace": "[t]he problem of organising a state, however hard it may seem, can be solved even for a race of devils, if only they are intelligent". Utterly self-interested *rational* devils would still realise that their long-term interests are best served by foregoing many of their short-term interests. Such devils would understand that seeking individual short term gain by looking to take benefits without paying costs - a little bit of cheating here and there - will ultimately defeat long term gain by reducing trust and therefore potential group size. Such cheating would ultimately lead to less trust and cohesion within the group that in consequence harms each individual. But the problem is that it *can* only be solved for a race of devils, or other such mythical beasts, because it requires a degree of rationality far beyond anything *Homo sapiens* is capable of. Reason cannot free us from culture any more than reason could have freed us from our genes. Over 100 years ago Huxley tried to argue that it had been man's reason that had originally freed him from the dictates of natural selection; an "apparent bifurcation ... as if man could somehow lift himself out of nature" (see Richards [1987]). Salvation from culture does not lie in reason, any more than salvation from genes did.

Of course you can simulate Iterated Prisoner's Dilemma on a computer; computers are perfectly logical, with a single binary programme. Even other animals display a greater degree of rational (by nature's standards) behaviour than our own species because they follow the dictates of a single very logical inheritance system set at conception. But man? Well man, for good or ill, has two inheritance systems. Two largely incompatible inheritance systems, and a second inheritance system that is being added to and re-written every second we live. And Kant's race of devils would have to be *continuously* and *perfectly* rational for large-scale cohesion to occur under our genetic inheritance system (which would otherwise "default" to or "settle" at the smaller group sizes actually found in nature), let alone under our cultural inheritance system (where contingent madness, as Nietzsche said, often seems the norm).

So for Maynard Smith the theoretical arguments as to whether perfectly rational beings could operate in infinite group sizes are for him only an academic game. Neither our genetic inheritance system nor our cultural inheritance system leads him to believe that reason has all that much to say about large-scale human group sizes. "The difficulty for any account of society that assumes that individuals behave rationally is partly that experimental psychologists find little support for such an optimistic view." For example, he says, people form conclusions on the basis of insufficient data but are then reluctant to change their minds even when confronted by disconfirming evidence. "It is also partly that the world is full of groups whose behaviour benefits neither the group as a whole nor the individuals that compose it. ... Such groups are often motivated by fanatical beliefs which, to outsiders, seem irrational and self-destructive" (in *Games, Sex and Evolution*). "[R]eal societies are not exclusively, or even mainly, rational", he wrote with Szathmáry. Dawkins shares Maynard Smith's understanding. After describing in *The Selfish Gene* the wonderful game theory computer simulations that have been run to demonstrate just what is possible if players continue to act rationally, he continues: "[s]adly, however, when psychologists set up games of Iterated Prisoner's Dilemma between real humans, nearly all players succumb to envy".

Yet even if humans are generally more rational in their daily lives than I give us credit for, the suggestion that our imperfect reason could ever be sufficient to free us from the dictates of cultural conditioning seems unrealistic. All mammals have differing capacities for reasoning. What higher capacities for reasoning tend to produce, however, are smaller group sizes. While less intelligent primates can live in larger groups, chimpanzees (canny buggers) can co-exist only in groups with a maximum size of around one hundred. The explanation for this is partly that, since they are more intelligent, the opportunities to successfully cheat the other group members grow. Groups of sixty to one hundred are therefore a stable group size that allows for close interaction while minimising the opportunities for undetected cheating. A larger brain does not necessitate any less cheating, only more intelligent cheating.

So what can we think about the autonomy of reason drawing us towards a "universal point of view"? The point of view of the Greeks perhaps, men better trained in logic than any of our contemporaries? Where Plato held that enemies captured in war should be exterminated? Enemies were barbarians; only Greek lives held value. Or Aristotle, who saw no incongruity in holding to the "objective" viewpoint that reason dictated that only non-Greeks should be slaves? The reason of the Middle Ages scholars perhaps, the great logicians of their time, who spent their days establishing wonderful justifications for Christian extermination of the Church's enemies? Or maybe the reason of Kant, that great champion of the "autonomy of reason"? Whose work, as Nietzsche realised, was often only a rational justification of Kant's own Christianity. Nietzsche wrote the following in Part One of *Beyond Good and Evil*, a section entitled "On the Prejudices of Philosophers": "[I]t is high time to replace the Kantian question 'how are synthetic judgements *a priori* possible?' with another question: 'why is belief in such judgements *necessary?*' – that is to say, it is time to grasp that, for the purpose of preserving beings such as ourselves, such judgements must be *believed* to be true; although they might of course still be *false* judgements!" Synthetic *a priori* knowledge for Kant meant profound truths about the world known independently of experience; necessary truths which became imperatives of action. And how is such knowledge possible?

"By means of a faculty" – he had said, or at least meant. But is that – an answer? An explanation? Or is it not rather merely a repetition of the question?

Friedrich Nietzsche – Beyond Good and Evil

As Edward Wilson himself admits in *On Human Nature*: "[I]f the mind is to any extent guided by Kantian imperatives, they are more likely to be found in religious feeling than in rational thought". Wilson then went on to argue that our religious codes could only be a product of genetics, not reason. Yet Darwinism tells us that such codes cannot be genetic in origin. And if you exclude reason, as Wilson does, contingent culture is all you have left.

The "faculty" that gave us the objective truths of reason, Kant said. The "faculty" resulting from evolution that gave us anti-Darwinian conscience and morality, the evolutionary psychologists say. The "faculty" of the mind that gave us devilishly perfect reason, the game theorists say. Why is belief in such things *necessary?* Nietzsche asked.

And on this very one-sided, very personal, view we shall leave reason to others. There is still much to sort out. But maybe, just maybe, there is still much human arrogance to overcome here too.

Artificial Life, Artificial Death

In the Spring of 2000 Hans Moravec and Ray Kurzweil, two of the leading lights in American artificial intelligence (or "A.I."), undertook a debate at Stanford University. The issue under discussion was the danger inherent in the development of A.I.

Moravec is filled with boundless optimism about the benefits of unrestrained A.I. research. He believes that humanness can be, first, replicated and, then, superseded by machines that will love their creators; that we can build what Moravec calls electronic "superhumanism". European A.I. has thrown up similar figures. Hugo de Garis, one of the biggest names in artificial brains, believes that A.I. will create "godlike supercreatures", "gargantuan" intelligences with mental capacities "trillions of trillions of times above our level", he writes. But where Moravec welcomes such a future, de Garis fears it; de Garis is convinced that these godlike supercreatures will come to view us as we view the lower primates, will even one day step on us as we step on ants.

But the A.I. auguries of hope and the A.I. auguries of doom both make the same mistake. They see the world-changing potential of artificial intelligence coming with the capacity for superhuman intelligence. But this is to totally miss the point of Huxley's "apparent paradox". Huxley's Paradox says that the danger within A.I. doesn't approach sometime in the far distant future when machines can have "godlike" intelligence. The danger exists from Day One because artificial intelligence can produce only *devils*, not gods.

Traditional A.I. grew out of the work of the codebreakers in World War Two who ended up designing vast machines to crunch information. After the war this led early A.I. practitioners to pursue a path of trying to simulate human intelligence using machines following a vast index of rules. Examples were the chess supercomputers designed to take on and beat the greatest grandmasters, like Deep Blue which beat Garry Kasparov in 1997.

The big limitation with this course of action was the need for huge lists of pre-programmed rules. This is not how natural life works, partly because it severely restricts innovative behaviour. So many computer scientists turned instead to the strategy of using simple components designed to interact to give rise to complex behaviour. This was the rise of the neural net; networks of artificial brain cells. Like animal brains, such brains can actually learn new things and recognise and evaluate simple patterns. They learn by trial and error.

But the problem with both the traditional approach to A.I. and the neural net approach is that they are difficult to apply to a wide range of situations. Being able to handle multiple novel situations is a crucial test of true intelligence, so a third approach developed. This third approach to A.I. doesn't just mimic animal brains, it mimics the evolutionary process that gave rise to animal brains. This was to be called artificial life, or A-Life, and the first American A-Life conference was held in 1987, while the first European conference was held in 1991. Robots are evolved by a process that imitates natural selection, where inchoate robots are allowed to compete, with the fittest being used to breed the next generation of robots, and so on.

As you can probably now appreciate, trying to mimic a brain that has been created by natural selection is dangerous enough. But trying to create intelligence by mimicking the process of genetic evolution is potentially lethal. Genetic evolution gives rise to pitiless indifference to the suffering of others, as any chimp infant or elephant seal pup will attest. Chimpanzees live in a world without compassion. Yet robot life created this way is potentially far more lethal than anything natural selection can create.

Earlier we considered the implications of orthodox evolutionary theory for the development of organic intelligent life. I argued that any extraterrestrial species could only have developed true intelligence and civilisation if it had been capable of escaping its primary replicator, of escaping the gross immorality of the genetic world. I then only briefly touched upon another question. Could natural life evolve intelligence and/or civilisation *without* the need to escape a primary replicator? Much human intelligence is a product of our capacity to form large groups, divide labour, and thus allow some the leisure to do theology, philosophy and science, to advance human thought and knowledge. But social insects can form huge groupings with division of labour without needing to escape their genetic programming; could such a pattern of life theoretically produce the basics of technological civilisation? In other words, would it ever be possible for *natural* life to remain subservient to the gross immorality of genetic nature, yet still evolve sophistication? I don't know; but I do know that it is a viable option for *artificially* created life.

Kant wrote: "[t]he problem of organising a state, however hard it may seem, can be solved even for a race of devils, if only they are intelligent". Artificial life, developed to the pattern of genetic natural selection, will produce machines that, like chimps, will be devils. Chimps cannot exist in large groupings, cannot develop complex division of labour, precisely because of the rules of genetic natural selection. However, unlike chimps, such machines will be perfectly rational devils. They will be able to follow the dictates of the gross immorality of a primary replicator yet will still be able to cohere in very large groups. Unlike natural world creatures, they won't follow their short-term interests, but only their long-term interests, and thus will be able to co-operate on a scale unknown in the natural world.

Artificial Life will have all the immorality of nature with none of its drawbacks. There will be nothing to stop such machines grouping together and acting against whatever built them. And the only action we can ever expect of them is the pitiless indifference of nature.

"the instinct of sympathy"

In 1996, a *Times* interviewer asked Dawkins why, if we are born selfish apes, are we so capable of being generous and sympathetic. "I ask him how we have managed to create minds which, although they reside in Darwinian brains, are so capable of being generous and sympathetic. 'Umm …' He pauses. I venture that I'd love to hear him say 'I don't know'. Eureka. 'I think I don't know,' he says with a ghost of a smile" (Dougary [1996]).

Dawkins does not mean that he doesn't know why we display unbelievable levels of certain types of "altruism". He is very happy with his 20-year idea that mental "viruses" play us for puppets, manipulating us first this way and then that. God-memes, nationalism-memes, ideology-memes, he points to them all. Other even more respected Darwinians (notably George Williams) have a great deal of time for this idea of manipulation by cultural parasites, and Darwin, too, saw how easily people are moulded by beliefs and by rituals. So whether we see such cultural characteristics as particulate agents invading susceptible human minds, or as nothing more than largely indistinguishable parts of a background cultural milieu, the power of these factors is not what is under discussion.

No; what Dawkins can't explain is not the ideologically-driven "altruism" of the suicide bomber, or the ritual-driven "altruism" of the Janissary, but the non-ideologically driven altruism of the unpaid, and truly loving, adoptive parent. At the end of his chapter on "memes" in *The Selfish Gene*, he comments cryptically that: "It is possible that *yet another* unique quality of man is a capacity for genuine, disinterested, true altruism. I hope so, but I am not going to argue the case one way or the other, nor to speculate over its possible memic evolution" (my emphasis). What on earth can he mean? What I have given you so far may not seem a particularly rosy image of what it means to be human; we are machines, manipulated this way and that by heartless forces that make us in their image. Now this is nothing to grieve over *per se*. These forces made possible all that you see around you. Be grateful that our early ancestors gibbered in terror at the power of their shamans (while always remembering the horrors that a grateful mystic will do for his shaman's gods). Had we not been susceptible to myth and ideology we would still be picking fleas off each other. Yet maybe there is something ... if not more, then at least a little more honourable. As I have written, I do not believe that reason truly allows us to cut the strings of the puppetmasters and move into a new realm of universal viewpoints; maybe I will be proved wrong, but that is not my concern right now. What I am interested in for the moment is a puppet that is capable of true love, and of real kindness.

What Dawkins is discussing as "true altruism" is empathy, or what Darwin, in *The Descent of Man*, called "the instinct of sympathy; and this instinct no doubt was originally acquired ... through natural selection". Yet we need an explanation for why such a uniquely self-abnegating capacity arose. This is so very far from the "blind, pitiless indifference" (Dawkins) of nature, yet still so far from "a driving purpose in life compatible with his self-interest" (Wilson) that allows for ideological re-writing, that we must be cautious. I am now going to suggest something that I have no real evidence for.

My suggestion is that a highly evolved capacity for simulation via self-inspection, directed by cultural forces (Dawkins' "memic" root above) might be one explanation for "genuine, disinterested, true altruism". The cultural root is still needed, because culture largely dictates to whom we will ever be truly altruistic. Even the wonderful people who adopt and love unrelated children still tend to limit the range of children they may consider adopting, and most adopt from within the same culture. Our love and our grief, too, tend similarly to be partially circumscribed.

First some background. What do I mean by "the capacity for simulation via self-inspection"? As Dawkins tells us in *The Selfish Gene*, the psychologist Nicholas Humphrey has spent much time studying other primates. In *The Inner Eye* Humphrey starts by telling us that he began wondering what the relatively large brains of, for example, gorillas were for. He started testing the hypothesis that they allowed such animals to deal effectively with the day-to-day social problems that living in tightly interacting groups throws up. As Dawkins writes: "A social animal lives in a world of others, a world of potential mates, rivals, partners, and enemies. To survive and prosper in such a world, you have to become good at predicting what these other individuals are going to do next". Humphrey had been one of the first to begin thinking along these lines: "Gorillas and chimps, I thought, had evolved to be psychologists by nature". To operate effectively in tight groups animals needed insight into the world of others, and somewhere along the line a new sense organ evolved which allowed this closer living. This new sense organ was the capacity for insight through simulation based on self-inspection, or the "inner eye": "We could, in effect, imagine what it's like to be them, because we know what it's like to be ourselves". So under Humphrey's hypothesis, the "inner eye" allows apes, and possibly some other relatively large-brained mammals, "a wolf-pack, say, a school of whales, a family of elephants", to operate together more efficiently. But make no mistake - this is still the selfish gene in operation. Such hypothetical simulation would have evolved purely because it makes the life of each individual a little easier. Natural selection can act only through and for the good of each being, as Darwin wrote. This capacity has nothing to do with decency or virtue. Such simulating primates still tear apart infants carrying competitors' genes. Social animals that may have this capacity for simulation are just as much programmed according to the gross immorality of nature as are all other less introspective natural world creatures. Infanticide and cannibalism are just as common. Simulation by itself does not escape the gross immorality of nature, it just makes that gross immorality function a little more efficiently. Other apes simulate, not so that they will know when to bring an ape feeling low a box of *Quality Street*, but so that they will know when to stay away, when it's safe to approach, and just how far to push any confrontation.

So this is Humphrey's hypothesis. Other Darwinians have also found it convincing, not least Richard Dawkins. "In his book, *The Inner Eye*, Humphrey makes a convincing case that highly social animals like us and chimpanzees have to become expert psychologists. ... Each animal looks inwards to its own feelings and emotions, as a means of understanding the feelings and emotions of others. ... [H]is book is persuasive". Dawkins though questions (as now does Humphrey) how far simulation by self-inspection can help us to understand the issue of consciousness itself. The relationship between simulation and consciousness is one that is today hotly debated by cognitive psychologists, but the two issues can be largely separated for our purposes. What I begin wondering is whether this (hypothesised) capacity for simulation by self-inspection may be the root of human empathy; something that does not, cannot, exist in the genetic world. I suggest that culture (usually acting through the very powerful forces of habit, ritual and so on) forces us, consciously and unconsciously, not just to occasionally simulate how another ape is feeling, but to live *continuously* within this "world of others". Continuously put in the position where we will "imagine what it's like to be them". And maybe then it's a small step from self-inspection as a means of understanding others to actually "seeing yourself" in another, to truly feeling another's pain, to living within their world.

A highly developed capacity for simulation, with simulation directed, enhanced and reinforced largely through inclusive, or semi-inclusive, belief systems and habits. This is all that I can think of to explain, for example, human adoption, where no one denies there can often be genuine love. We certainly aren't genetically capable of it (it is not, as Dawkins told us earlier, an evolutionarily stable strategy), and any answer of "well, we just can" rather misses the point. We are apes. We are born selfish apes. So we need to know *why* we can adopt. It's not genetic, so it must be "exogenetic". One form of exogenetic reprogramming occurs through ideological manipulation. Is another form of exogenetic reprogramming perhaps operating through simulation?

To stress again, the hypothesised primate capacity for simulation, *if* it is being modified to work in the ways that I suggest, does not on its own make us decent. Simulation, no matter how highly evolved, is a tool of natural selection. Simulating chimpanzees will tear apart an unrelated baby chimpanzee faster than you can tear paper, and with the same level of indifference. Simulating humans still have this same infanticidal genetic programme that first needs to be overcome by cultural reprogramming. And simulating humans reprogrammed through ideology can still butcher Jewish or Midianite babies. But the capacity for simulation, acted on by cultural influences resulting from close living – itself originally made possible by ideological manipulation - and acted on by the ideological manipulation itself, may have made kindness possible under certain circumstances and in certain specific directions. This is purely an hypothesis, and one that may be difficult to test. It is easy to prove genetics cannot be responsible for the great part of human behaviour simply because we know the rules of natural selection; it will, however, be a lot harder to resolve exactly how culture manipulates the genetic. Yet there are many pressing reasons for trying to establish what role (if any) highly developed primate simulation may be playing. Understanding such paradigmatically human characteristics as love and grief may depend on our answer. In chapter 7 we saw the error of trying to claim that adaptation can explain grief. But so far we have only seen that the evolutionary psychologists cannot explain grief, and not that cultural manipulation can. So without simulation – or something similar - the model of man may be incomplete.

Genes – natural selection - cannot explain grief. And grief as purely a product of manipulation, of ideology, ritual and myth, seems unconvincing. Oh, I don't deny that there are many instances of ideological grief. When there is beating of breasts and much wailing and gnashing of teeth when one of "our boys" goes down behind "enemy" lines, the ideology just drips out. But even here there may be simulation partly in play. We still feel for the young soldier, we still live in his world for a brief moment, even if it is our ideologies that have told us to feel for him in the first place. Correspondingly our ideologies tell us when not to feel empathy, because someone is not one of "our" boys, or is unholy in the sight of our God. Yet the sort of grief a parent feels for a dying or long dead child, or that we can feel for the death of a very close friend, seems to me to have no real element of manipulation to it. I think that because of this primate capacity for simulation, enhanced by the close contact of cultural living, humans can spend so long living inside another's skin, of concerning themselves with another's hopes and dreams, that when that person dies it is as if they really were losing a part of themselves.

Love, like grief, seems to require some additional explanation. Genetics cannot explain love; the love for a partner, a close friend, an adopted child. Genetics cannot even explain the love of a mother for her natural child (nature, remember, is about sacrifice, not love. "Motherly love" for a weak infant in the genetic world is the crunch of mum's jaws). There is no love in the natural world. Some animals do take the unusual step of partnering for life but *not* because they have "fallen in love"; selfish gene theory and observation tell us that they stay together simply because evolution has found this the best route for genetic success. Any of these animals would be long gone if it found a way of making its partner care for their young in a secure environment. It would be hightailing it out of there without a backward glance. And because there is no love in nature, the sociobiologists' explanations of love for a human partner are amusing to say the least. On the BBC Radio 4 "Basic Instincts" production first mentioned in chapter 3 Robert Wright explained that love had evolved in man as a form of *self-deception*. There is "love" only because man, uniquely, gets deceived by nature into thinking he's committed to partners of the opposite sex. "The man deceives *himself* into thinking he's committed to the woman, thus convinces *her* of that and winds up having sex", he suggested. Brings a lump to the throat, doesn't it? Yet strangely this "efficient strategy" is seen nowhere else in nature. Neither is love for offspring, yet evolutionary psychologists barely bother to come up with "Just So" Stories explaining kin-directed love, such is their utter misunderstanding that a grossly immoral process could ever produce ordinary human morality. And in chapter 7 we met similar group-selectionist theories in respect of love for adopted children.

We only have a finite number of options. Neither love nor Wright's "love-substitute" exist in nature, or could exist in nature given that self-deceit is a characteristic that, even if it could get started, would tend to leave an animal vulnerable to being taken advantage of. In four billion years nature did not produce love, or love-substitute, for the very good reason that it is wasteful and inefficient, and would leave a creature at the mercy of less decent conspecifics. If love is not genetic, it must be exogenetic. Is love a simple end product of peer pressure, ritual and ideology? Or is love perhaps something rather different; total empathic understanding of another made possible by cultural forces acting on our capacity to simulate?

Much more thought needs to be given to what is going on in most instances of empathy, and specifically such anti-Darwinian capacities as love and grief. It is important to note here that philosophers have always tended to use empathy and simulation as an indicator that something extraordinary is occurring in the higher mind far removed from the butcher's yard of natural selection or the madness of culture. For philosophers self-awareness has been the key to explaining such distinctly human attributes. In *Beyond Evolution: Human Nature and the Limits of Evolutionary Explanation*, the philosopher Anthony O'Hear discusses Jean-Paul Sartre's idea of "the gaze", "something not taken into account by the functionalist accounts of the likes of Dennett". "The gaze of another self-conscious being ... pulls me up short, makes me conscious of my own self, and at the same time conscious of the self who is not me." Central to Sartre's concept of the gaze was the sense of shame it could bring with it. It should be remembered that Darwin himself saw shame as one of the most powerful factors in the development of the moral conscience. As he wrote in *Descent*: "But there is another and much more powerful stimulus to the development of the social virtues, namely, the praise and the blame of our fellow-men", and Darwin saw this "love of approbation and the dread of infamy" as having arisen primarily from the capacity for sympathy.

O'Hear, then, is not denying the materialist root of the gaze, only that adaptation designed the brain to react to others in the way it does. This is the "emergent properties" argument we touched upon in an earlier chapter, or an example of what hierarchical theorists term "exaptation" so as to indicate structures not originally built by natural selection (that is, not arising adaptively) but subsequently co-opted. I have reservations that self-consciousness in itself (like simulation in itself) can explain much given that other intelligent apes also have a fair degree of self-consciousness. However, our *level* of self-consciousness and our apparently considerable capacity to simulate may be one of the keys to taking a selfish cannibal and turning it into an altruistic human. Mindful of the preceding qualification, O'Hear summarises quite nicely: "What I have been suggesting over the last few pages is that one evolutionary advantage of self-consciousness for human beings may be that they are able to predict each other's behaviour better and so be more successful in the competition and co-operation of social life. But, once established, this type of mutual fellow feeling leads to demands for reciprocity unknown in the animal kingdom". Self-consciousness looks for recognition and validation from other self-consciousnesses, says O'Hear. "The personal in us cannot be satisfied by the impersonal, but requires recognition by other persons. In self-consciousness we move from the mechanical, even though flexible, operation of desire and its gratification to one's existence being regarded as a source of value".

One final point about simulation which should be considered separately. Whether or not I am correct in suggesting that simulation can begin to induce true kindness, Nick Humphrey's hypothesised capacity for simulation may itself be one of the explanations for why we are so susceptible to cultural manipulation in the first place. It is often said that man is the only animal with knowledge of his own mortality. Man is also the only animal with knowledge of his own individual insignificance in the great cosmic scheme of things. Simulation – including the capacity to think ourselves into extinction and pointlessness - may make facing these sorts of truth so frightening and uncomfortable that the mind seeks solace in comforting illusions and ideologies that tell us about life after death, or the special place in history for certain peoples.

No apologies
This chapter of four essays is very different from the rest of the book. The rest of the book can be proved: the four billion year genetic code of nature, the "delusion" of free will, the implications for the "me" that could exist separate from genes and culture, and the contingent insanity this seems to deed us. In contrast, this chapter stands alone, suggesting ideas that I cannot be certain about, though I do believe each of the essays to be both scientifically acceptable and logically necessary. But I make no apologies if parts of this chapter are less useful (and ultimately less valid) than other parts of this book. Poor speculation is a little better than no speculation at all, providing the problems being addressed are explained and reservations are clearly spelt out. One of the crucially important points that this book has to make is that, so far, no one has really tried to understand the mind of the human animal through

Darwinian theory. Because of this we are going to have so many issues to resolve within human Darwinism when we at last begin to apply gene-selectionist understandings to ourselves. When we at last begin to see ourselves as (biologically) *apes*, as *selfish apes*.

11

Huxley's Paradox

Simple cannibalism ... can be expected in all animals except strict vegetarians.
George C. Williams – "Huxley's Evolution and Ethics"

We are animals and, as Darwin realised, human morality can thus only be explained as culture overwriting biology. Contrary to the anti-Darwinian propaganda of the human sociobiologists, the behavioural geneticists and the evolutionary psychologists, man is genetically just another ape. The pattern of the natural world is, as George Williams says, "gross immorality", and it is our "enemy". Our ancestors never broke this template of nature and evolved morality; we are a product of natural selection, and products of natural selection carry savage murder in their genes. We are born cannibal. We must learn to be human.

For those looking for someone to blame for the last twenty-five years of heretical human Darwinism, don't blame Darwin. He did his best. It was not Darwin who "lets us down". It was we who let Darwin down.

Quis custodiet?

"I will nail him out like a kite to a barn door, an example to all evil doers", Thomas Huxley wrote in 1862 as he was preparing to publish *Evidence as to Man's Place in Nature*. He was writing about Richard Owen, the great nineteenth-century palaeontologist, backbone of the Christian establishment, and enemy of both Darwin and Darwinian evolution. Owen represented the forces of scientific conservatism, hostile, if not necessarily to evolution, then at least to natural selection; to an evolution that invoked a blind, physical process. The Huxley-Owen clash would emerge over the most furious evolutionary argument of the period - what became known as the "hippocampus question". This debate embodied all of the fundamental questions that were raised by Darwin's work: was man a product of the natural world, of billion-year-old blind physical forces, or was he perhaps something more divine? It was at this point that Disraeli would famously ask: "Is man an ape or an angel?"

Owen, the doyen of British palaeontology, believed that man wasn't just a primate with a very large brain, but that we had special attributes that must have been bequeathed us by our Creator. Owen held that proof of this could be found in brain structure; that the human brain possessed a "unique lobe", the hippocampus minor, not found in a gorilla's brain. Man should be classified apart from the rest of nature, Owen claimed, because he was as different from a chimpanzee as the ape was from a platypus. Huxley, however, believed differently. In October 1862, in front of the British Association for the Advancement of Science, Huxley would dissect an ape brain to reveal the hippocampus, and Owen's career was effectively over. As Adrian Desmond notes in his biography of Huxley, Darwin's Bulldog "was smashing the post-Waterloo consensus which made man sacrosanct". But today Darwinism needs its Thomas Huxleys perhaps more than at any other time in its history. It needs them to fight this arrogant belief that, in one giant leap 100,000 or so years ago, man freed himself from the dastardly clutches of nature. It needs them to fight the Richard Owens of today; otherwise respected scientists and theorists protecting the popular wisdom of man as innately better than the rest of the natural world.

While much of the fanciful biological speculation of the last few decades can be put down to simple naivety some is harder to excuse. Biology will always suffer to an extent from the prejudices of scientists, and as George Williams noted sadly in *Adaptation and Natural Selection*: "Perhaps biology would have been able to mature more rapidly in a culture not dominated by Judeo-Christian theology and the Romantic tradition". When we have "Darwinians" seeking to put us outside orthodox evolutionary theory we should all begin to look a little closer. When we have "Darwinians" who celebrate free will it is time to start asking serious questions.

I treat free will as a sort of marker for initial evolutionary orthodoxy and proficiency. Free will is a concept that has no place in a Darwinian worldview, a worldview where nothing gets to live outside the causal universe. Free will is, as Darwin said, a delusion. Coming from a political philosophy background, I have less of a problem with social scientists and other non-Darwinians who wish to debate the existence of free will. It is fascinating, if a little unnerving, to watch the logical summersaults social scientists attempt in its defence. While I was at LSE I remember coming across a major constitutional theorist who was perhaps a little more intellectually honest than most who get deeply involved in the free will debate. He simply stated baldly that he believed in free will because he "couldn't do political theory" if he didn't believe it existed[37]. If free will does not exist, he thought, then what is the point? The point, of course, is that we are what we are and we must learn to live with what we have, and that what we have is really not that bad. But while I can still have some time for a social scientist who wishes to keep free will in the picture, I have grave reservations about "Darwinians" who extol the virtues of free will. Darwinism is, in its very essence, a materialist science, so it is rather strange to seek to attach to it non-materialism. Yet the problem with evolutionary theory, and genetic theorising in general,

---

[37] This was not the most frightening defence of free will I came across at LSE. A well-published free will theorist explained to me that it must exist because otherwise two brothers who had had "the same upbringing" would act identically. My jaw actually dropped. Even behavioural geneticists have stopped arguing that any two humans can ever have had "the same upbringing". The evolutionary psychologist Robert Wright put it quite nicely in *The Moral Animal*: "The larger point here is about 'nonshared environment', whose importance geneticists have grasped only over the last decade". Wright explains that though two brothers do share some aspects of their environment (same parents, same school) a large part of their environment is "nonshared" (relationships with parents, siblings, teachers and friends, etc.).

is that so often peer review fails us, for reasons we touched upon in this book. And until we get effective peer review, Darwinism will continue to need its bulldogs, and kites must expect to be nailed out.

The philosopher of science Philip Kitcher's 1985 *Vaulting Ambition* was a detailed 450-page exposition of the numerous flaws of early sociobiology. While today a little dated in its minutiae (and perhaps relying a little too easily on hierarchical interpretations), many of Kitcher's logical, mathematical, observational and methodological criticisms are as valid today as they were over fifteen years ago. Kitcher's programme had been to show that while properly conducted biological investigations of human behaviour are to be welcomed, loose and inaccurate speculations are not. "[T]he defects lie in the method, not the matter. Wilson might have brought to his study of human behavior the same care and rigor that he has lavished on ants. But he did not." Kitcher showed that while non-human biology was usually characterised by careful, well-reasoned and well-tested hypotheses, human Darwinism often did little more than shine a spotlight on the theorists themselves. When biologists fail to stamp out bad theorising, when they look away while their colleagues bask in the popular acclaim that all too often follows bad evolutionary theorising, people get hurt. Those whose aspirations have been destroyed and whose lives have been reduced through the application of misguided science direct us to look closely at any theorising that might lead us to further mistakes, Kitcher wrote. "Their descendants deserve better."

Vaulting Ambition was published more than fifteen years ago, and it is largely thanks to the efforts of philosophers of biology like Kitcher, as well as to numerous "hierarchical" biologists beginning with Gould and Lewontin, that human sociobiology was largely abandoned, to be replaced by the (on the surface) much more careful reasoning of evolutionary psychology. Policing biology was a duty that some outside gene-selectionism did not eschew (even if their motivations perhaps sometimes had more to do with a commitment to opposing theories than a noble desire to protect Darwin's legacy). Yet today Darwinism needs a policeman more than at any other time in the past. Today people are still being hurt by bad Darwinism, for giving a wrong answer in human biology is usually worse than giving no answer at all. Some evolutionary psychologists are as keen as the early human sociobiologists to give an evolutionary account of xenophobia. Yet evolution cannot explain racism or xenophobia; evolution cannot even explain why we can sit peacefully in a cinema full of strangers for Heaven's sake. We are not "in a sense the ants of the ape family". We are, in every sense, apes, we are the apes of the ape family, and apes cannot co-exist in groups of larger than one hundred or so.

In *Vaulting Ambition* Kitcher noted that: "Many of the critics of pop sociobiology write as if there ought to be a single crucial flaw in the program. Criticism would be simpler if that were so. ... Unfortunately, sociobiology is a motley. ... There is a family of mistakes, and in any distinct examples distinct members are implicated." Sociobiology, both father and son, is indeed a motley. We have seen human sociobiologists, including Edward Wilson himself, who sought to explain both human similarities and most human differences in genetic terms. Other human sociobiologists, notably Richard Alexander, largely sought only to show that human similarities are genetic, but have then tried to argue that human behaviour can be understood purely in terms of continual maximisation of inclusive fitness. Today we have evolutionary psychologists who reject those earlier attempts; they dismiss the accounts that tried to explain human differences in terms of evolution, and they argue loudly that we should be seen as the executors of 100,000-year-old adaptations, not as modern day fitness maximisers. Yet, too, we have evolutionary psychologists like Steven Pinker who also wish to put much human difference down to genetics, and trumpet the "uncanny" findings of behavioural geneticists like Thomas Bouchard. *Plus ça change*. And every theorist has pursued his or her own models, his or her own reasoning, his or her own mistakes. Yet Kitcher is so very wrong to think that all these pop sociobiologists, these evolutionary heretics, do not at the same time share a single crucial flaw. It is a flaw that bound Wilson to Alexander and Barash, and today binds these forerunners to Pinker, Cronin, Cosmides and Daly, as well as to their legions of devoted followers in science writing and the media, such as Gribbin, Wright and Matt Ridley. It is a flaw that starts from the assumption that we are not animals, that humans do not share the genetic code of the rest of the natural world.

Make no mistake, it is not up to Darwin, to Huxley, to Williams or to Maynard Smith to prove to you that we carry the same genetic code as the rest of nature. You, like all chimpanzees, carry the genetic code for cannibalism. The theory of evolution by natural selection makes this conclusion inevitable, makes this conclusion the automatic "default assumption" of modern biology. Anyone who argues against Darwin, against Wallace and Dawkins, and suggests that evolution caused us to answer to a different genetic tune must prove his or her case before the eyes of the scientific community. The scientific community owes it to Darwin to be ready to evaluate the argument this time - it cannot simply eschew a potentially unpleasant task, pretend to be a little hard of hearing, otherwise engaged in the coming decades. Philosophers of biology and social scientists should not need to police Darwinism; that should be for biologists to do. And if you wish to gainsay Darwin and Wallace, your argument cannot be based on Disney-esque anthropomorphisms, or gut feel, or that shiver you get down your spine when carefully observing natural world behaviour. It cannot be based on the tears of Sarah Hrdy. Human Darwinism, if it is to mean anything at all, must be more than what Kitcher calls yet another "opium for the people": "those who lull the conscience of the people by supplying palatable palliatives are less easily pardoned."

I consider it the greatest tragedy that Stephen Jay Gould was not a selfish gene theorist. Gould, like Bishop Wilberforce in the nineteenth century, was all too aware that Darwinism will always attract the most unwelcome enthusiasts. As Dennett has written: "Probably no area of scientific research is driven by more hidden agendas than evolutionary theory". Had Gould used his formidable learning and his gift for writing in the services of selfish gene-ery (rather than so often against it) the last twenty-five years might have been so very different. It is time selfish gene theorists and hierarchical theorists realised that the greatest threat to the Darwinism they both love comes not from each other, and not from those who simply dismiss Darwinism - it comes from those who would twist the theory to fit a worldview. It comes from those who are "implicit" in their theorising, not those who are explicit. And for all their differences, both traditions generally have the courage to be explicit. Until the two traditions begin to patch up their differences, although we will all be losers, it will be Darwin who will remain the greatest loser of all.

"the Huxley-Williams nature-as-enemy idea"
George Williams calls it the "Huxley-Williams nature-as-enemy idea", Alfred Russel Wallace called it "the inverse problem", but Thomas

Huxley simply called it that "apparent paradox" of evolution by natural selection: "that ethical nature, while born of cosmic nature, is necessarily at enmity with its parent." Philosophers have largely given up the fight for the soul of Darwinism. But Darwinism is truly worth fighting for. It is what philosophers have been searching for for two and a half thousand years, it is nothing less than the philosophers' stone. By letting us know one side of the equation, we get to understand the other side. Darwinism gives to modern philosophers the tool to prove many of the insights of so many great philosophers, from Plato and Aristotle onwards. It changes base observations into golden wisdom by the singular process of allowing us to fix one of the two variables. As some philosophers are starting to realise, when we set aside myths like free will, we have only two variables to consider, genes and culture. Hugely complex variables, of course, almost infinitely divisible, but, crucially, with at least one of them according to a set pattern. A pattern that is in many ways predictable. This is partly why genetics will always be so strongly fought over, because those who understand the gene have inherited the Earth by the process of elimination. He who controls interpretation of the gene largely controls interpretation of culture. As Maynard Smith reminds us, "[f]or a geneticist, all variance which is not genetic is, by definition, 'environmental'". The argument inevitably turns on the nature of the genetic programme itself. A programme that, despite Darwin's detailed instructions, we have so poorly understood.

The social sciences first turned their backs on Darwinism as it was being presented because they had seen it all before. Social scientists, knowledgeable about history, know that we are the genetic issue of people who had taken part in horrific slave trading a few hundred years ago. The offspring of those who had butchered and tortured with glee in the Wars of Religion, and those who had raised rape, infanticide and theft to the very keystone of culture in pre-Christian times. Throughout our long, dark history our forefathers have been capable of extraordinarily savage behaviour. And so when some people began again to claim that such acts in *individuals* were the result of genetic programming, while such acts in *peoples* were usually (but not always) seen as the result of cultural programming, social scientists knew better.

Yet Darwinism, true Darwinism, never claimed any of this. True Darwinism teaches us the immense power of cultures to mould the human character. True Darwinism teaches us that we, like all animals, have an innate nature, but that it is a nature with no goodness, no virtue and no honour. Yet it is also a nature with no hatred, no racism, and (excepting social insects) no extermination. All that is best about us, and almost all that is worst, comes from the power of culture to twist the human brain. Darwin realised that human morality could only be understood as culture acting on biology. Today, when both observation and selfish gene understandings teach us about the true extent of the horror in nature - knowledge of which Darwin was largely unaware - this conclusion screams itself from the rooftops. We did not break from the pattern of nature, because the pattern that we observe in nature *is* nature. Selfishness is not a by-product of natural selection, it is the engine driving natural selection. There is no "outside" we can get to. Reciprocity does not "overcome" selfishness because in nature reciprocity *is* selfishness, just another trick used to further the ends of a grossly immoral system. For the reasons we have discussed, natural selection could not ensure that mankind conformed to a different pattern. This is our genetic inheritance, whether we like it or not. This is our innate nature, but this is not us. *From* the brutes, but not *of* them, as Huxley put it. The answer to the human animal is the answer that the greatest philosophers have given across the ages - culture makes us what we are. We are no *tabula rasa*, but what is *subsequently* written onto one very special newborn ape makes it what it will become. Darwin gave us this knowledge.

A most precious gift
Darwin gave us the most precious gift of all. He gave us the chance to grow up. He gave us the chance to understand ourselves, and in so doing a chance to escape the murderous insanity of our past. Throughout our history we have based our moral codes on just about everything imaginable; everything, that is, except the truth about ourselves. We have listened to bigots, fools, monsters and misguided populists. We have flocked to anyone who will make us feel proud and better about ourselves, irrespective of whether we had anything to be proud of.

To you it may not be a big deal that Moses exterminated the Midianites, killed all their pregnant women, killed all their babies and male children, killed or enslaved all their female children. "It was a long time ago ... forget about it." But I cannot forget about it. In that genocidal slaughter lay the seeds of all future atrocities, both large and small. I am not talking about historical connections, of course, but psychological connections. In the hypocritical barbarity of Moses, the atrocities which "lit up hell-fires in Christendom" as Darwin wrote, lay the pattern for so much human thinking and action. But in his acts lie also the seeds of the *explanation* for what Nietzsche called our "madness". When we consider Moses, we begin to perceive the Nazis who exterminated Jews and Slavs, the Communists who liquidated the kulaks, the sectarian bomber in Northern Ireland placing a device in a crowded street, even the vicious hypocrisy of the modern Californian. I cannot "forget about it" because it is central to the gift Darwin gave us - the chance to understand ourselves, and, in understanding, a chance to escape from the horror. Every act of religious, or political, or nationalistic, intolerance has its psychological counterpart in those seven days three thousand years ago. Darwin taught us about the one animal programmed through two fundamentally opposing languages, one language grossly immoral, but the other a language of contingency and doublethink. That there is no "us" to choose separate from our murderous genes and our often all-too-bigoted cultures. If we are to save ourselves, our hope lies not in gods, or "free will", or little green men from Mars. Our salvation, our true salvation, will lie only in the knowledge of our actions that Darwinism offers us.

Yes, Darwinism is new territory, and yes, it raises serious concerns. That, of course, makes it frightening as well as valuable. But Darwinism does allow us to keep the best of our heritage, while abandoning only the worst. Nevertheless this is largely academic, because, for better or worse, truth is like the fabled genie of the lamp. The genie is often used as a metaphor for the new genetic knowledge, but the metaphor works better than most people imagine. It works not only as an allegory about the power and autonomy of new knowledge, but also as a good allegory for its (mis)use. For many, many years, the stories tell us, the genie is kept prisoner in the lamp by wicked rulers who use him to serve their nefarious purposes. Rulers who pass him from father to son to maintain an empire based on the rule of might, not right, and cruelly mistreat those they see as beneath them. But then, one day, some poor innocent rubs the lamp by accident and makes a deal that sets the genie free. And then it's impossible to get the little blighter back into his lamp. Once he gets a taste of freedom he's off like a rat up a drainpipe (to mix my metaphors). But not until he has traditionally brought freedom and hope to those who suffered under the wicked rulers. So, whether or not you relish the opportunity Darwin had deeded you, evolutionary theory is going to start becoming a liability to those who have traditionally twisted it to serve their own interests.

Darwinism is a science; the problem is that we forgot this for a while. We allowed Darwinism to become the plaything of a select few. Darwinism is science, and as such it will no longer serve the interests of the few. In 1827 an eighteen-year-old Darwin looked on as vested interests saw the danger of a new scientific order and sought to ban it. Others sought to twist it to their ends. Such abuse and misuse of science must end. If you think that Darwinism will continue to lie for you, think again. If you think that Darwinism will lie to protect your vicious hypocrisy, you are deluding yourself. There is no free will; to blame is to be unjust, as Darwin and Nietzsche both wrote. And you were born just another selfish ape. As Huxley put it, "the foundation of morality is to have done, once and for all, with lying; to give up pretending to believe that for which there is no evidence". In his biography, *Naturalist*, Edward Wilson writes that: "There is no finer sight on green Earth than a defeated bully". After eleven chapters, I suddenly find myself in agreement with this human sociobiologist.

And a gift worth having

Being a Darwinian is sometimes hard. To know that each of us has only one life to live; one brief span of years; one short flicker in the eternal darkness. Knowing this is not particularly hard. What is hard is to stand silently by while some people speak of a better life to come, and of resigning oneself to misery now in anticipation of a later existence. Because there is no such afterlife, no reincarnation, no second chance. All that each of us has or will ever have is encompassed by the years we have here and now. Perhaps this makes you mad. Perhaps the thought that there is no greater purpose, no chance to make amends, no chance for it all to come right, makes you angry. That some children will die senselessly from disease, and hunger, and neglect, and that they will not be comforted in the arms of a loving God. I sure hope it makes you mad. It makes me mad as hell. It makes me want to do everything I can to improve life here and now.

Huxley's Paradox

Some people fear Darwinian knowledge. To me it is liberating. It makes us realise that we must help ourselves, that no god (or, after the last chapter, extraterrestrial) is going to magically appear to help us. Darwinism does not kill gods, but it does kill the idea of an interventionist god. Darwinism has room, if you so wish, for a god who set the universe in motion with the full realisation that intelligent life would one day evolve. With the realisation that that life would do untold horrors, and kindnesses, to its own. With the realisation that intelligent life would only ultimately find salvation in knowledge. Darwinism has room, perhaps, for a creator god who wept quiet tears for the stupidity he foresaw for that one species, and eagerly anticipated the day when this species realises that its answers lie in truth, not fiction. That, although wonderfully kind men like Jesus Christ have given us hope, their hope has always been fleeting and unpredictable, because their messages can so easily be twisted or misunderstood. A god who foresaw that true hope for man lies in science, not superstition. Perhaps Darwin is even a prophet of this god.

I have no desire to necessarily see a continued struggle between science and religion for the simple reason that science has so much to thank religion for. Superstition first turned an ape into a human, and even today religion is capable of such immense goodness. Reason alone may never be enough for man, and I would rather people took refuge in a watered-down religion based on universal kindness than in perhaps some more murderous and divisive political or nationalistic ideology. But science and religion can only come to an accommodation when religion accepts scientific knowledge. Free will does not exist, and man is born just another ape. And there can only be an accommodation when religion stops teaching hatred and intolerance.

But what about those of us who do not believe in any form of god? Is our world-view any less agreeable? I find the Darwinian view both breathtaking and liberating. To accept the contingencies that took another murderous ape and created all that you see around you; that is truly a story as wondrous as any ever told. To understand that we are masters of our destiny, that we *make* purpose, that with every generation we get to write what "human" means. Not because of some silly belief in "free will", but because we understand that we can change the world from what it is today, no matter that each of us is a product of what went before. Determinism in no way reconciles one to fatalism.

Human history has been a quest for civilisation, a slow (and unsteady) move away from barbarism. Our future is not locked in our unchanging genes; the future is up to us, for better or worse.

And maybe after all there is an afterlife. Each of us does live on after his or her death. Not through some imagined soul, and not through some naïve faith in genes. You have better things to impart to your descendants than cannibalistic urges. No; each of us will live on in the ripples we leave in life. In the effects we have on the lives we touch during our time on this planet. We live on in the deeds we do and the effect those deeds have on others. Culture makes us, but culture is made up of a multitude of individual lives. By touching those other lives we change forever those lives, and those changes will be transmitted to future generations when those lives touch yet more lives. Some lives we change in small ways and some we change immeasurably. The parent and the friend, the teacher who makes the difference, the scientist who finds a way to ease suffering, the aid worker who relieves distress, even the rare politician who cares more for truth than ideology or honours. We are like rocks dropped in a very large lake. As the waves move outward in all directions gradually the energy dissipates and the waves get smaller. But by the time those waves have become mere ripples the surface area they have covered is enormous. This, I feel, is an afterlife worth honouring.

"Let us understand, once for all ..."
It is a serious mistake "to assume that the objectivity of a science depends on the objectivity of the scientist. ... The natural scientist is just as partisan as other people", wrote the philosopher of science Karl Popper in his *27 Theses*. Science has an unfortunate history of failing to think the unthinkable because science is never detached from the people who study it. When Darwin brought forth his idea of sexual selection in 1871 - that sexual attractiveness is a key force acting in evolution - it contained two very different strands: male-male competition and female choice. Most of Darwin's (exclusively male) scientific contemporaries could accept the former answer but not the latter, because females did not choose in Victorian England. Even through to the 1960s female choice was ignored in biology, and it was only with the rise of feminism and the female entry into higher levels of biology that female choice started to be wholeheartedly embraced by

natural scientists, and today forms one of the key areas of evolutionary research. What such histories teach are that: 1. Only the best of scientists are capable of thinking "outside the box" of prejudice and received opinion, and 2. Science often moves forward when some have the courage to reject the ideologies of their contemporaries.

The last word, then, must go to Huxley. At the end of the day, this is not so much a book about Darwinism as a book about what has gone so badly wrong in modern human Darwinism. In his essay "Biological Potentiality vs. Biological Determinism", Stephen Jay Gould tells us that the great eighteenth-century taxonomist Carolus Linnaeus was uncertain of how to classify his own species to complete the definitive edition of his *Systema Naturae*. Would he rank man as just another animal, or would he create for us a distinct status? Linnaeus compromised, says Gould. He placed us within his classification (close to monkeys and bats), but then set us apart by his description. He defined our relatives by mundane characteristics such as size and shape, but "[f]or *Homo sapiens*, he wrote only the Socratic injunction: *nosce te ipsum* - 'know thyself.' For Linnaeus, *Homo sapiens* was both special and not special."

And that really is the crux of the matter. No one who has studied nature in any detail doubts that we are special. The question is what is it that makes us special? Darwin, Wallace, Huxley, Williams, Maynard Smith and Dawkins all join (indeed, surpass) the hierarchicals in realising that we are special because something is freeing us from our biology. If we discount Wallace's view that some supernatural force has interfered with biology we are left only with the materialist answer of the other five - culture is overwriting, and often completely re-writing, biology. Yet for the last twenty-five years we have had a second option presented to us. Wilson, Pinker, and Cronin, to name but three, believe that in producing man nature threw away the pattern it has used since life originated on this planet. They argue that nature completely re-wrote both the rules of evolutionary inheritance and our ancestors' genetic code in creating us. Man, the 2,000,000,000-to-1 experiment. Man, who got to break a four billion-year mould. Man, the grossly immoral ape that evolved morality and virtue. Man, blessed child of a beneficent "Mother" Nature.

The tragedy is not that these have been "the Darwin Decades" (as they have been described), the tragedy is that they have not been. The tragedy is not that we have no biologist worthy to inherit Darwin's mantle, the tragedy is that we have no one to continue Huxley's work. Where once Darwin found bulldogs, today he finds only lap-dogs. Darwin's memory, and incomparable legacy, deserves far better than the spectacle it has become; of anodyne evo-babble, of ceaseless attempts to pretend that we are not products of the natural world. It is not Darwin who, as Dr. Cronin puts it, "lets us down". For the last twenty-five years it is we who have let Darwin down by uncritically accepting such arrogant creeds. The natural world is a world of gross immorality. It is a world of species-wide coding for cannibalism, infanticide, extreme violence and pitiless indifference. This is a world that laughs at virtue, decency, honour, duty, love and hate. Contrary to the empty speculations of the last twenty-five years of vocal human Darwinism this is not a world we out-evolved. This is a world we can never out-evolve. This is a world each one of us carries in the nucleus of every cell in his or her body. This is a world that exists in every twist of the double helix. This is nature, in all her glory.

This is our innate nature, but this is not "us". As Thomas Huxley put it, man might be *from* the brutes but "he is assuredly not *of* them". Like selfish gene biologists today, Huxley and Darwin both realised that it is culture that frees us from the billion-year mould of nature. It is culture that both saves and damns us. It is culture that makes us human.

Let us understand, once for all, that the ethical progress of society depends, not on imitating the cosmic process, still less in running away from it, but in combating it.

Thomas H. Huxley – 1893

Afterword

Is it still science when we refuse to ask the difficult questions because we are afraid of the answers we might get?

For Darwin, human self-knowledge was the highest and most interesting problem for the naturalist, and it is one and a half centuries since he taught us that we are products of evolution by natural selection. It is almost four decades since George C. Williams showed us that natural selection operates at the level of the gene. And yet in all that time, no one has had the courage to stand up and loudly ask one very obvious question: *Surely humans must carry the same genetic code as the rest of nature?*

I have spent the last five years trying to get people to ask this question. I have written to journals, newspapers, science journalists, science editors and television presenters. When the Nuffield Council on Bioethics grandly announced in late 2000 that it had set up a working party to establish the veracity of claims made in behavioural genetics, I asked it to consider the Huxley-Williams nature-as-enemy idea. I even asked *Nature* - the all-important science journal set up by Huxley, and opened by him with Goethe's words: "Nature! Everyone sees her in his own fashion" - to ponder Huxley's Paradox. But no one seemed to want to hear. This is a book I never expected or wanted to write, but felt both a moral and intellectual duty to do so. *Surely man must accord to the genetic pattern found in the rest of nature?* So obvious a question, but a frightening question, a question with such profound implications. Implications for genetics, philosophy, law, religion, ethics, society and the nature of the scientific process.

Afterword

Many find life without free will "almost too grim to contemplate". A century and a half after Darwin, just as many scientists would seem to find the human genetic code almost too grim to contemplate.

I think I shall avoid the whole subject, as so surrounded with prejudices, though I fully admit that [man's origin] *is the highest & most interesting problem for the naturalist.*

Charles Darwin – (letter to A. R. Wallace, 1858)

Letter
G. C. Williams to the author, December 1998
(edited with permission)

STATE UNIVERSITY OF NEW YORK

Division of Biological Sciences
Department of Ecology and Evolution

3 December 1998

I sure thank you for sending "Unnatural selection" and for its kind words about me. I found the whole thing well conceived and presented, and most enjoyable. I can make only a few critical comments, arising not so much from disagreement as from a desire to help you avoid some ill conceived but likely criticisms.

It may be unwise to say that group selection "will not occur" (page 596). Something like "usually too weak to produce noteworthy effects" might be better. D S Wilson (no relation to E O) and others have produced evidence of group selection producing female-biased sex ratios in species with special population structures. More recently, people (even I) have been saying that parasite virulence will reflect the balance between within-host and between-host selection. Further down on that page

...although perhaps you are already too obsessed with your Darwinian triumvirate. The last chapter in the book elaborates on the Huxley-Williams nature-as-enemy idea. The *Altruism* piece is for the MIT Encyclopedia of Cognitive Science, which should be out next spring some time.

Thanks again, and I wish you the best of luck in your work, and in all else.

Yours truly

George C Williams
Professor Emeritus

Bibliography

Alexander, R.D. [1987]: *The Biology of Moral Systems*, Hawthorne, NY, Aldine de Gruyter.
Allen-Mills, T. [1996]: 'Motherly gorilla saves injured boy at Chicago zoo', *Sunday Times*, 18 August, Pt. 1, p. 15.
Arens, W. [1980]: *The Man-eating Myth: Anthropology and Anthropophagy*, New York, Oxford University Press.
Aristotle [c.4th B.C.]: *Ethics*, trans. J.A.K. Thompson, (1976) London, Penguin.
Aristotle [c.4th B.C.]: *The Politics*, trans. T. A. Sinclair, (1981) London, Penguin.
Badcock, C. [1991]: *Evolution and Individual Behaviour: An Introduction to Human Sociobiology*, Oxford, Basil Blackwell.
Barash, D. [1979]: *Sociobiology: The Whisperings Within*, (1980) London, Souvenir Press.
Barash, D. [1982]: *Sociobiology and Behavior* (expanded edition to first edition published 1977), New York, Elsevier.
Barkow, J.H., **Cosmides, L.** and **Tooby, J.** (*eds*.) [1992]: *The Adapted Mind: Evolutionary Psychology and the Generation of Culture*, New York, Oxford University Press.
Barrett, P.H., **Gautrey, P.**, **Herbert, S.**, **Kohn D.** and **Smith, S.** (*eds*.) [1987]: *Charles Darwin's Notebooks, 1836-1844: Geology, Transmutation of Species, Metaphysical Enquiries*, transcribed and edited by Paul H. Barrett and others, Cambridge, Cambridge University Press.
Bennett, N.C. and **Faulkes, C.G.** [2000]: *African Mole-Rats: Ecology and Eusociality*, Cambridge, Cambridge University Press.
Berry, A. [2000]: 'An ugly baby' (a review of *Footsteps in the Forest: Alfred*

Bibliography

Russel Wallace in the Amazon by Sandra Knapp), *London Review of Books*, 18 May, pp. 26-7.

Boyd, R. and **Richerson, P.J.** [1985]: *Culture and the Evolutionary Process*, Chicago, University of Chicago Press.

Bradie, M. [1994]: *The Secret Chain: Evolution and Ethics*, New York,: State University of New York Press.

Brockman, J. (*ed.*) [1995]: *The Third Culture: Beyond the Scientific Revolution*, New York, Simon & Schuster.

Brockman, J. (*ed.*) [2000]: *The Greatest Inventions of the Past 2000 Years*, London, Phoenix.

Brown, A. [1999]: *The Darwin Wars: How Stupid Genes Became Selfish Gods*, London, Simon & Schuster.

Buckland, W. [1835]: 'On the discovery of coprolites, or fossil faeces, in the lias at Lyme Regis, and in other formations', *Transactions of the Geological Society of London*, 2nd Series, Pt. 3, pp. 223-36.

Bugental, D.B., **Blue, J.** and **Cruzcosa, M.** [1989]: 'Perceived control over caregiving outcomes: implications for child abuse', *Developmental Psychology*, **25**, pp. 532-9.

Bygott, J.D. [1972]: 'Cannibalism among wild chimpanzees', *Nature*, **238**, pp. 410-11.

Campbell, D.T. [1975]: 'On the conflicts between biological and social evolution and between psychology and moral tradition', *American Psychologist*, **30**, pp. 1103-26.

Christman, J. (*ed.*) [1989]: *The Inner Citadel*, Oxford, Oxford University Press.

Confucius [c.5th B.C.]: *The Analects*, trans. D. C. Lau, (1979) London, Penguin.

Cronin, H. [1991]: *The Ant and the Peacock: Altruism and Sexual Selection from Darwin to Today*, Cambridge, Cambridge University Press.

Daly, M. and **Wilson, M.** [1988]: *Homicide*, Hawthorne, NY, Aldine de Gruyter.

Darwin, C. [1859]: *On the Origin of Species by Means of Natural Selection; or The Preservation of Favoured Races in the Struggle for Life* (edited and introduced by J. W. Burrow), (1968) London, Penguin.

Darwin, C. [1871]: *The Descent of Man, and Selection in Relation to Sex* (facsimile reproduction of first edition with an introduction by J. T. Bonner and R. M. May), (1981) Princeton, NJ, Princeton University Press.

Davies, P. [1989]: 'The new physics: a synthesis'. In P. Davies (*ed.*) *The New*

Bibliography

Physics, (1996) Cambridge, Cambridge University Press, pp. 1-6.

Dawkins, R. [1981]: 'In defence of selfish genes', *Philosophy*, **56**, pp. 556-73.

Dawkins, R. [1982]: *The Extended Phenotype: The Gene as the Unit of Selection*, (1983) Oxford, Oxford University Press.

Dawkins, R. [1983]: 'Universal Darwinism'. In D.S. Bendall (*ed.*) *Evolution from Molecules to Men*, Cambridge, Cambridge University Press, pp. 403-25.

Dawkins, R. [1986]: 'Sociobiology: the new storm in a teacup'. In S. Rose and L. Appignanesi (*eds.*) *Science and Beyond*, Oxford, Basil Blackwell, pp. 61-78.

Dawkins, R. [1986a]: *The Blind Watchmaker*, Harlow, Longman.

Dawkins, R. [1989]: *The Selfish Gene* (revised edition to first edition published 1976), Oxford, Oxford University Press.

Dawkins, R. [1996]: 'How we got a head start on our animal natures', *The Sunday Times*, 29 December, section 3, p. 6.

Dawkins, R. [1998]: *Unweaving the Rainbow: Science, Delusion and the Appetite for Wonder*, London, Allen Lane.

Dennett, D.C. [1984]: *Elbow Room: The Varieties of Free Will Worth Wanting*, (1996) Cambridge, Massachusetts, The MIT Press.

Dennett, D.C. [1993]: 'Confusion over evolution: an exchange', *New York Review of Books*, 14 January, pp. 43-4.

Dennett, D.C. [1995]: *Darwin's Dangerous Idea: Evolution and the Meanings of Life*, (1996) London, Penguin.

Dennett, D.C. [2003]: *Freedom Evolves*, London, Allen Lane.

Desmond, A. [1997]: *Huxley: From Devil's Disciple to Evolution's High Priest*, (1998) London, Penguin.

Desmond, A. and **Moore, J.** [1991]: *Darwin*, (1992) London, Penguin.

de Waal, F.B.M. [1996]: *Good Natured: The Origins of Right and Wrong in Humans and Other Animals*, Cambridge, Massachusetts, Harvard University Press.

de Waal, F.B.M. [1997]: *Bonobo: The Forgotten Ape*, Berkeley, California, University of California Press.

de Waal, F.B.M. [1998]: '"The social behavior of chimpanzees and bonobos: empirical evidence and shifting assumptions": Reply', *Current Anthropology*, **39**, pp. 407-8.

Diamond, J. [1991]: *The Rise and Fall of The Third Chimpanzee*, (1992) London, Vintage.

Dougary, G. [1996]: 'Gene genie', *The Times magazine*, 31 August, pp. 8-13.

Einstein, A. [1954]: *Ideas and Opinions*, (1982) New York, Three Rivers

Bibliography

Press.

Eldredge, N. [1995]: *Reinventing Darwin: The Great Evolutionary Debate*, (1996) London, Phoenix.

Elgar, M.A. and **Crespi, B.J.** [1992]: 'Ecology and evolution of cannibalism'. In M.A. Elgar and B.J. Crespi (*eds*.) *Cannibalism: Ecology and Evolution Among Diverse Taxa*, Oxford, Oxford University Press, pp. 1-12.

Fortey, R. [2000]: *Trilobite! Eyewitness to Evolution*, London, HarperCollins.

Freud, S. [1927]: 'The future of an illusion', trans. J. Strachey. In A. Dickson (*ed*.), *Sigmund Freud: Civilization, Society and Religion*, (1991) London, Penguin, pp. 183-241.

Gibbard, A. [1990]: *Wise Choices, Apt Feelings: A Theory of Normative Judgment*, Oxford, Clarendon Press.

Gould, S.J. [1977]: *Ever Since Darwin: Reflections in Natural History*, (1991) London, Penguin.

Gould, S.J. [1992]: 'The confusion over evolution', *New York Review of Books*, 19 November, pp. 47-54.

Gould, S.J. [1993]: 'Confusion over evolution: an exchange', *New York Review of Books*, 14 January, p. 44.

Gould, S. J. [1993a]: *Eight Little Piggies: Reflections in Natural History*, (1994) London, Penguin.

Gould, S.J. [1996]: *Dinosaur in a Haystack*, (1997) London, Penguin.

Gribbin, M. and **Gribbin, J.** [1993]: *Being Human: Putting People in an Evolutionary Perspective*, (1995) London, Phoenix.

Griffiths, P.E. [1995]: 'The Cronin controversy', *The British Journal for the Philosophy of Science*, **46**, pp. 122-38.

Haldane, J.B.S. [1955]: 'Population genetics', *New Biology*, **18**, pp. 34-51.

Hamai, M., **Nishida, T.**, **Takasaki, H.** and **Turner, L.A.** [1992]: 'New records of within-group infanticide and cannibalism in wild chimpanzees', *Primates*, **33**, pp. 151-62.

Hamilton, W.D. [1964]: 'The genetical evolution of social behaviour, II'. In W.D. Hamilton (ed) *Narrow Roads of Gene Land: The Collected Papers of W. D. Hamilton, Volume I: Evolution of Social Behaviour*, (1996) Oxford, W.H Freeman, pp. 47-81.

Hamilton, W.D. [1971]: 'Selection of selfish and altruistic behaviour in some extreme models'. In W.D. Hamilton (ed) *Narrow Roads of Gene Land: Volume I*, (1996) Oxford: W.H Freeman, pp. 198-227.

Hamilton, W.D. [1975]: 'Innate social aptitudes of man: an approach from evolutionary genetics'. In W.D. Hamilton (ed) *Narrow Roads of Gene Land:*

Volume I, (1996) Oxford, W.H Freeman, pp. 329-351.
Hiraiwa-Hasegawa, M. [1992]: 'Cannibalism among non-human primates'. In M.A. Elgar and B.J. Crespi (*eds.*) *Cannibalism: Ecology and Evolution Among Diverse Taxa*, Oxford, Oxford University Press, pp. 323-38.
Hopkin, K. [1999]: 'The greatest apes', *New Scientist*, **162**, 15 May, pp. 26-30.
Horgan, J. [1993]: 'Eugenics revisited', *Scientific American*, **268**, June, pp. 122-31.
Horgan, J. [1995]: 'The new social Darwinists', *Scientific American*, **273**, October, pp. 150-7.
Horgan, J. [1996]: *The End of Science: Facing the Limits of Knowledge in the Twilight of the Scientific Age*, (1997) London, Little Brown & Company.
Hrdy, S.B. [1977]: 'Infanticide as a primate reproductive strategy', *American Scientist*, **65**, pp. 40-9.
Hrdy, S.B. [1977a]: *The Langurs of Abu: Female and Male Strategies of Reproduction*, (1980) Cambridge, Massachusetts, Harvard University Press.
Humphrey, N. [1986]: *The Inner Eye*, (1993) London, Vintage.
Humphrey, N. [1995]: *Soul Searching: Human Nature and Supernatural Belief*, London, Chatto & Windus.
Huxley, A. [1932]: *Brave New World*, (1964) London, Penguin.
Huxley, T.H. [1863]: *Evidence as to Man's Place in Nature*, (1959) University of Michigan Press.
Huxley, T.H. [1869]: 'Nature: aphorisms by Goethe', *Nature*, **1**, pp. 9-11.
Huxley, T.H. [1886]: 'Science and morals'. In T.H. Huxley *Collected Essays: Volume IX*, (1968) New York, Greenwood Press, pp. 117-46.
Huxley, T.H. [1894]: *Evolution and Ethics: and Other Essays*, (1901) London, Macmillan.
James, O. [2002]: *They F*** You Up*, London, Bloomsbury.
Jones, J.S. [1982]: 'Of cannibals and kin', *Nature*, **299**, pp. 202-3.
Jones, J.S. [1994]: *The Language of the Genes: Biology, History and the Evolutionary Future*, London, Flamingo.
Jones, J.S. [1996]: *In the Blood: God, Genes and Destiny*, (1997) London, Flamingo.
Jones, J.S. [1997]: 'The set within the skull', *New York Review of Books*, 6 November, pp. 13-6.
Jones, J.S. [1999]: *Almost Like a Whale:* The Origin of Species *Updated*, London, Doubleday.
Kalb, C. [2000]: 'Drugged-out toddlers', *Newsweek*, 6 March, p. 62.
Kano, T. [1998]: '"The social behavior of chimpanzees and bonobos: empirical

evidence and shifting assumptions" : Reply', *Current Anthropology*, **39**, pp. 410-11.

Kant, I. [1785]: *Foundations of the Metaphysics of Morals*, trans. Lewis White Beck, (1985) London, Collier Macmillan.

Kant, I. [1795] 'Perpetual peace', trans. Lewis White Beck. In Lewis White Beck (*ed.*) *Kant Selections*, (1988) London, Collier Macmillan, pp. 430-57.

Kitcher, P. [1985]: *Vaulting Ambition: Sociobiology and the Quest for Human Nature*, Cambridge, Massachusetts, The MIT Press.

Kymlicka, W. [1990]: *Contemporary Political Philosophy: An Introduction*, Oxford, Clarendon Press.

Lewontin, R.C. [1994]: 'Women versus the biologists', *New York Review of Books*, 7 April, pp. 31-35.

Lumsden, C.J. and **Wilson, E.O.** [1981]: *Genes, Mind and Culture: The Coevolutionary Process*, Cambridge, Massachusetts, Harvard University Press.

Malik, K. [2000]: *Man, Beast and Zombie: What Science Can and Cannot Tell Us About Human Nature*, London, Weidenfeld & Nicolson.

Maynard Smith, J. [1964]: 'Group selection and kin selection', *Nature*, **201**, pp. 1145-7.

Maynard Smith, J. [1988]: *Games, Sex and Evolution*, Hemel Hempstead, Harvester Wheatsheaf.

Maynard Smith, J. [1993]: 'Confusion over evolution: an exchange', *New York Review of Books*, 14 January, p. 43.

Maynard Smith, J. [1996]: 'Conclusions'. In W.C. Runciman, J. Maynard Smith and R.I.M. Dunbar (*eds.*) *Evolution of Social Behaviour Patterns in Primates and Man: A Joint Discussion Meeting of the Royal Society and the British Academy*, (and published as *Proceedings of The British Academy*, **88**), Oxford, Oxford University Press, pp. 291-7.

Maynard Smith, J. and **Szathmáry, E.** [1995]: *The Major Transitions in Evolution*, Oxford, W. H. Freeman.

Medawar, P. B. [1977]: 'Pro bono publico', *The Spectator*, 15 January, p. 20.

Medawar, P. B. [1977a]: 'Unnatural Science', *New York Review of Books*, 3 February, pp. 13-18.

Midgley, M. [1979]: 'Gene-juggling', *Philosophy*, **54**, pp. 439-58.

Miles, J.B. [1998]: 'Unnatural selection', *Philosophy*, **73**, pp. 593-608.

Miles, J.B. [2000]: 'Darwin's final message: we have no honour', *Children & Society*, **14**, pp. 110-20.

Milton, K. [1998]: '"The social behavior of chimpanzees and bonobos:

Bibliography

empirical evidence and shifting assumptions": Reply', *Current Anthropology*, **39**, pp. 411-12.

Mineau, P. and **Cooke, F.** [1979]: 'Rape in the Lesser Snow Goose', *Behaviour*, **70**, pp. 280-91.

Nietzsche, F. [1878]: *Human, All Too Human*, trans. M. Faber and S. Lehmann, (1994) London, Penguin.

Nietzsche, F. [1886]: *Beyond Good and Evil*, trans. R.J. Hollingdale, (1990) London, Penguin.

Nietzsche, F. [1892]: *Thus Spoke Zarathustra*, trans. R.J. Hollingdale, (1961) London, Penguin.

O'Hear, A. [1997]: *Beyond Evolution: Human Nature and the Limits of Evolutionary Explanation*, Oxford, Oxford University Press.

O'Riain, M.J., **Jarvis, J.U.M.** and **Faulkes, C.G.** [1996]: 'A dispersive morph in the naked mole-rat', *Nature*, **380**, pp. 619-621.

Paradis, J. and **Williams, G.C.** [1989]: *Evolution & Ethics: T.H. Huxley's Evolution and Ethics With New Essays on its Victorian and Sociobiological Context*, Princeton, New Jersey, Princeton University Press.

Pascal, B. [ca. 1660]: *Pensées: Notes on Religion and Other Subjects*, trans. J. Warrington, (1960) London, J.M. Dent & Sons.

Pereboom, D. [2001]: *Living Without Free Will*, Cambridge, Cambridge University Press.

Petrinovich, L. [2000]: *The Cannibal Within*, Hawthorne, New York, Aldine de Gruyter.

Pinker, S. [1994]: *The Language Instinct: The New Science of Language and the Mind*, (1995) London, Penguin.

Pinker, S. [1997]: *How the Mind Works*, (1998) London, Penguin.

Pinker, S. [1997a]: 'Evolutionary psychology: an exchange', *New York Review of Books*, 9 October, pp. 55-6.

Pinker, S. [2002]: *The Blank Slate: The Modern Denial of Human Nature*, London, Allen Lane.

Plato [c.4th B.C.]: *The Republic*, trans. D. Lee, (1974), London, Penguin.

Pollack, R. [1995]: *Signs of Life: The Language and Meanings of DNA*, London, Penguin.

Radford, T. [1995]: 'Message from the forest', *The Guardian*, 31 August.

Richards, R.J. [1987]: *Darwin and the Emergence of Evolutionary Theories of Mind and Behavior*, Chicago, University of Chicago Press.

Ridley, Mark and **Dawkins, R.** [1981]: 'The natural selection of altruism'. In J.P. Rushton and R.M. Sorrentino (*eds.*) *Altruism and Helping Behavior:*

Bibliography

Social, Personality and Developmental Perspectives, Hillsdale, NJ, Lawrence Erlbaum, pp. 19-39.

Ridley, Matt [1993]: *The Red Queen: Sex and the Evolution of Human Nature*, (1994) London, Penguin.

Ridley, Matt [1996]: *The Origins of Virtue*, Harmondsworth, Viking.

Ridley, Matt [1999]: *Genome: The Autobiography of a Species in 23 Chapters*, London, 4th Estate.

Robertson, I.H. [1999]: *Mind Sculpture: Your Brain's Untapped Potential*, London, Bantam Press.

Robinson, D.N. [1996]: *Wild Beasts & Idle Humours: The Insanity Defense from Antiquity to the Present*, (1998), Cambridge, Massachusetts, Harvard University Press.

Ruse, M. [1989]: *The Darwinian Paradigm: Essays on its History, Philosophy, and Religious Implications*, London, Routledge.

Sagan, C. [1983]: *Cosmos*, London, Abacus.

Segerstrale, U. [2000]: *Defenders of the Truth: The Sociobiology Debate*, Oxford, Oxford University Press.

Sibley, C.G. and **Ahlquist, J.E.** [1984]: 'The phylogeny of the hominoid primates, as indicated by DNA-DNA hybridization', *Journal of Molecular Evolution*, **20**, pp. 2-15

Singer, P. [1981]: *The Expanding Circle: Ethics and Sociobiology*, (1983) Oxford, Oxford University Press.

Sober, E. and **Wilson, D. S.** [1999]: *Unto Others: The Evolution and Psychology of Unselfish Behavior*, Cambridge, Massachusetts, Harvard University Press.

Stanford, C.B. [1998]: 'The social behavior of chimpanzees and bonobos: empirical evidence and shifting assumptions', *Current Anthropology*, **39**, pp. 399-420.

Symons, D. [1992]: 'On the use and misuse of Darwinism in the study of human behavior'. In Barkow and others (*eds.*) *The Adapted Mind*, pp. 137-159.

't Hooft, G. [2002]: 'Determinism beneath quantum mechanics', SPIN-2002/45, ITP-UU-02/69, quant-ph/0212095. E-print at http://xxx.lanl.gov/abs/quant-ph/0212095.

Thornhill, R. and **Palmer, C.T.** [2000]: *A Natural History of Rape: Biological Bases of Sexual Coercion*, Cambridge, Massachusetts, The MIT Press.

Tooby, J. and **Cosmides, L.** [1992]: 'The psychological foundations of culture'. In Barkow and others (*eds.*) *The Adapted Mind*, pp. 19-136.

Trigg, R. [1982]: *The Shaping of Man: Philosophical Aspects of Sociobiology*, Oxford, Basil Blackwell.

Trigg, R. [1988]: *Ideas of Human Nature: An Historical Introduction*, (1993) Oxford, Basil Blackwell.

Turner, J. [1999]: 'The inhuman gene?', *The Times Literary Supplement*, 19 February, pp. 5-6.

Urmson, J.O. [1988]: *Aristotle's Ethics*, Oxford, Blackwell.

van Inwagen, P. [1983]: *An Essay on Free Will*, Oxford, Oxford University Press.

van Noordwijk, M.A. and **van Schaik, C.P.** [2000]: 'Reproductive patterns in eutherian mammals: adaptations against infanticide?' In C.P. van Schaik and C.H. Janson (*eds.*) *Infanticide By Males and Its Implications*, Cambridge, Cambridge University Press, pp. 322-360.

van Schaik, C.P. [2000]: 'Vulnerability to infanticide by males: patterns among mammals'. In C.P. van Schaik and C.H. Janson (*eds.*) *Infanticide By Males and Its Implications*, Cambridge, Cambridge University Press, pp. 61-71.

van Schaik, C.P., van Noordwijk, M.A. and **Nunn, C.L.** [1999]: 'Sex and social evolution in primates'. In P.C. Lee (*ed.*) *Comparative Primate Socioecology*, Cambridge, Cambridge University Press, pp. 204-40.

Wallace, A.R. [1864]: 'The origin of human races and the antiquity of man deduced from the theory of "natural selection"', *Anthropological Review and Journal of the Anthropological Society of London*, **2**, pp. 158-187.

Wallace, A.R. [1891]: *Natural Selection and Tropical Nature: Essays on Descriptive and Theoretical Biology*, (1969) Farnborough, Gregg International.

Wilberforce, S. [1860]: 'On the Origin of Species, by Charles Darwin', *Quarterly Review* (reviewed anonymously), **108**, pp. 225-264.

Williams D.E. [1989]: *Truth, Hope and Power: The Thought of Karl Popper*, Toronto, University of Toronto Press.

Williams G.C. [1966]: *Adaptation and Natural Selection: A Critique of Some Current Evolutionary Thought*, (1996) Princeton, New Jersey, Princeton University Press.

Williams, G.C. [1988]: 'Huxley's Evolution and Ethics in sociobiological perspective', *Zygon*, **23**, pp. 383-407.

Williams, G.C. [1996]: *Plan & Purpose in Nature*, (1997) London, Phoenix.

Wilson, E. O. [1975]: *Sociobiology: The New Synthesis*, Cambridge,

Bibliography

Massachusetts, Harvard University Press.
Wilson, E. O. [1975a]: 'Human decency is animal', *The New York Times Magazine*, 12 October, pp. 38-50.
Wilson, E. O. [1978]: *On Human Nature*, Cambridge, Massachusetts, Harvard University Press.
Wilson, E. O. [1996]: *Naturalist*, London, Penguin.
Wind, J. [1984]: 'Review of some books on human sociobiology'. In J. Wind (*ed.*) *Essays in Human Sociobiology: Volume 1*, (1985) London, Academic Press, pp. 149-158.
Wright, L. [1997]: *Twins: Genes, Environment and the Mystery of Human Identity*, London, Weidenfeld & Nicolson.
Wright, R. [1994]: *The Moral Animal: Evolutionary Psychology and Everyday Life*, (1996) London, Abacus.
Wright, R. [1999]: 'The accidental creationist: why Stephen Jay Gould is bad for evolution', *The New Yorker*, 13 December, pp. 56-65.

Index

Adaptation and Natural Selection, 25-6, 115, 198, 221
Adelphophagy, 66
Adoption, 114-7, 190
Agnosticism, 24, 138
Ahlquist, Jon, 53, 220
Alcoholism, 103, 125
Alexander, Richard, 35, 200, 213
Altruism
    kin selection, 16, 18-9, 34, 42, 47-9, 76-8, 88, 218
    manipulation, 76
    reciprocal altruism, 14-6, 18-20, 28, 34, 35, 42, 49, 76-8, 194, 203
    technical, 15, 19
Arens, William, 51, 213
Aristotle, 1-2, 37, 80-1, 173, 183, 202, 213, 221
Artificial intelligence
    A.I., 185-6
    A-Life, 185-7
    race of devils problem, 185, 187
Artificial life. *See* Artificial intelligence
Atheism, 105, 138, 163
Attention-deficit hyperactivity disorder (ADHD), 118-25, 131, 133

Badcock, Christopher, 78, 213
Barash, David, 35, 114, 200, 213
Beetle, carrion, 64
Behavioural genetics, 39, 103-9, 120, 130, 144, 156
Bennett, Nigel, 44, 213
Bentham, Jeremy, 153
Berry, Andrew, 22, 213
Bonobo, 42, 53, 68-71, 127, 215, 217, 218, 220
Bouchard, Thomas J., 104-7, 200
Boyd, Robert, 28, 214
Bradie, Michael, 91-2, 214
Bragg, Melvyn, 104, 119, 144
British Association for the Advancement of Science, 139, 197
British Journal for the Philosophy of Science, The, 29, 45, 85, 216
Britten, Roy, 53
Brockman, John, 25-6, 153, 214
Brown, Andrew, 85, 89, 91
Browne, William, 157
Buckland, William, 66, 214
Bugental, Daphne, 121, 214
Burgess Shale, 87
Burt, Cyril, 125
Bygott, J. David, xi, 67, 214

Campbell, Donald, 172, 214

Index

Cannibalism, xi, xii, xiii, 34, 50-2, 54, 57, 59-62, 65-6, 68-9, 71, 73-5, 113, 127, 133, 189, 196, 200, 209, 214, 216, 217
Chamberlain, Neville, 98
Chambers, Robert, 3
Charity, 62, 63
Chimpanzee, xi, xii, 4, 12, 15, 32-4, 40, 42, 50, 53-4, 61, 63-4, 67-73, 113, 121, 124, 127, 130, 134, 166, 180, 183, 186-7, 189, 190-1, 197, 200, 214, 215, 217, 218, 220
China, 1995 eugenics law, 98
Chomsky, Noam, 5
Churchill, Winston, 98
Confucius, 80
Contingent ape, the, 175
Conway Morris, Simon, 87
Cooke, Fred, 62, 72, 219
Cosmides, Leda, 11, 200, 213, 220
Crespi, Bernard, 66, 216, 217
Cronin, Helena, 1-2, 20-1, 24, 45, 49, 73, 77, 78, 85-6, 88, 90, 94, 112, 137, 200, 208-9, 214, 216
Cuckoo, 76-8, 115
Cystic fibrosis, 56

Daly, Martin, 64, 113-5, 128, 134-5, 200, 214
Damned Man Walking hypothesis. *See* Free will: Darwin's Wager
Darwin Wars, The, 85, 89, 214
Darwin, Charles, xii, xiii, 1, 3-5, 8, 11, 15, 17-24, 26, 31, 35, 37, 39, 45, 51, 58-60, 63, 73-4, 79-80, 82-3, 88, 90, 95-6, 98-100, 103, 111, 116, 120-1, 134, 138-140, 142, 146, 148-9, 154-7, 158, 166, 169, 172-4, 181, 187-9, 193, 196-209, 211, 213-6, 219, 221
Darwin, Erasmus, 3
Darwin's Bulldog. *See* Huxley, Thomas H.
Darwin's Wager. *See* Free will: Darwin's Wager
Darwinism. *See* Natural selection
Darwinism *lite*, 156
Davenport, Charles B., 100
Davies, Paul, 143, 214
Dawkins, Richard, xiii, 5, 7-8, 13, 18, 20, 24-5, 29-31, 34, 38, 42, 45-8, 51, 56, 59, 62, 68, 71, 74-80, 82, 86-9, 91-3, 100, 114, 127, 140, 163, 165, 167, 172, 175, 177-9, 181, 183, 187-90, 201, 208, 215, 219
de Garis, Hugo, 185
de Waal, Frans, xii, 48, 53, 61, 70-1, 215
Deep Blue, 185
Dennett, Daniel, 11, 38-9, 49, 58, 63, 82, 86, 88, 90, 94, 96, 111, 115-7, 134, 146-9, 151-3, 156-8, 166, 174, 193, 201, 215
Depression, 103, 125
Descent of Man, The, 1, 20, 74, 79, 120, 188, 214
Desmond, Adrian, 23-4, 138, 157, 166, 169, 197, 215
Diamond, Jared, 53-4, 215
Diderot, Denis, 160-1
Disraeli, Benjamin, 197
Dobzhansky, Theodosius, 106
Dog, domestication of, 72
Doublethink, explanation of, 166, 170, 204
Dyslexia, 131

Index

Einstein, Albert, 158, 215
Eldredge, Niles, 5, 25-6, 83, 216
Elgar, Mark, 66, 216, 217
Emergent properties, 76, 193
Eugenics, 98-101, 217
Eugenics Record Office (US), 100
Evo-babble, 209
Evolution and Ethics, 59, 61, 79, 179, 196, 214, 217, 219, 221
Evolution by natural selection. *See* Natural selection
Evolutionarily stable strategy (ESS), 28, 32, 114, 177, 190
Evolutionary evangelism, 7, 8, 34
Evolutionary heresy, 37–58
Evolutionary psychology, 1, 4, 9-12, 16, 20, 26, 35, 39, 40-2, 46, 48-52, 54-8, 60, 64, 72-8, 82, 85, 87-9, 97, 100, 104, 106, 109, 111-7, 120, 127-30, 134-5, 137, 146-7, 180, 184, 191-2, 196, 199-200
Exaptation, 193
Extraterrestrials
   made in the image of man, 176-9, 186, 206
   universal Darwinism, 177, 215

Faulkes, Chris, 44, 213, 219
Firefly, 77
Fish, cleaner, 19
Fisher, R.A., 25
Fortey, Richard, 87, 216
Fossey, Dian, 67
Frasier, 116
Free will
   as anti-Christian belief, 155
   chaos theory and, 141, 143
   compatibilism, 148, 152

consequentialism, 153, 154
Darwin's Wager, 158-62
delusion of, 137-49, 150, 152, 154, 194, 198
Duracell definition of, 148
Noble Lie and, 154
quantum theory and, 142-3
unpredictability and, 141, 143-4, 148
Freud, Sigmund, 75, 180, 216

Galton, Francis, 98
Game theory, 27, 32, 78, 179, 183-4
Gene, definition of, 25
Genetic drift, 84
Genic selection. *See* Natural selection
Genocide, 167-71, 173, 175, 178, 204
Gibbard, Allan, 91, 216
Gibbon, 54
Gillie, Oliver, 125
Goethe, Johann Wolfgang von, 210
Goldschmidt, Richard, 57
Gombe National Park, 68, 69
Goodall, Jane, 68
Goose, snow, 62, 72
Gorilla, 53, 67, 70, 115, 189, 197
Gould, Stephen Jay, 8-9, 25-6, 46, 49, 60, 75, 83, 87-90, 96, 99, 100-1, 116-7, 142, 180, 199, 201, 208, 216, 222
Gribbin, John, 6, 12-3, 16, 47-9, 78, 138, 200, 216
Grief, 6, 62-3, 73, 111-2, 115-6, 122, 134, 150, 165, 189, 191-3
Griffiths, Paul, 45, 85-7, 216
Gross immorality of nature. *See* Cannibalism, Infanticide, Rape
Group selection. *See* Natural selection
Group size, 32-3, 42-3, 79, 182, 183

225

Index

Gull, blackheaded, 47

Haldane, J.B.S., 18, 25, 126, 216
Hamai, Miya, 68, 216
Hamer, Dean, 102, 128, 145
Hamilton, William D., xiii, 17-8, 43, 48, 61, 82, 84, 86, 89, 216
Harris, Judith Rich, 120
Heresy. *See* Evolutionary heresy
Heterocannibalism, 66
Heterodoxy. *See* Evolutionary heresy
Hierarchical theory. *See* Multilevel selection
Hippocampus question, the, 197
Hiraiwa-Hasegawa, Mariko, xii, 68, 217
Hobbes, Thomas, 92
Homo erectus, 54
Homo habilis, 54
Homophobia, 129
Horgan, John, 5, 9-10, 12, 103, 113, 129, 153, 217
Hrdy, Sarah Blaffer, xii, 63-5, 67, 72, 129, 201, 217
Human Behavior and Evolution Society (HBES), 129
Human sociobiology, 5-13, 35, 39, 50, 71, 75, 81, 87, 89, 92-3, 96, 116, 138, 199-200, 213, 215, 218, 220, 221, 222
Hume, David, 144, 148, 153, 180
Humphrey, Nicholas, 75, 189
Huxley, Aldous, 118
Huxley, Thomas H., xii, xiii, 15, 19, 21-4, 26, 31, 34, 45, 48, 59, 61, 73-4, 79, 92, 118, 134-5, 139-40, 151, 155, 163, 166, 172, 179-82, 185, 196-7, 200, 202-3, 205, 208-10, 215, 217, 219, 221

Huxley's Paradox, xii, xiii, 12, 23, 27, 45, 48, 74, 79-80, 135, 138, 141, 173, 175, 181, 185, 196, 201, 210
Huxley-Williams nature-as-enemy idea. *See* Huxley's Paradox
Hymenoptera. *See* Social insects

Immigration Restriction Act (US), 100
Individual selection. *See* Natural selection
Infanticide, xii, 34, 40, 52, 54, 57, 60-2, 64-6, 68-71, 73-4, 81, 113-8, 122, 128, 133-5, 170, 189, 202, 209, 217, 221
Institute of Psychiatry, 130
Inverse problem (Wallace), xii, 22-3, 201
IQ, 100, 125-6

James, Oliver, 120, 217
Jim Twins, The, 104-8
Jones, Steve, 62, 67, 72, 91, 103, 119, 124-5, 132, 144, 217
Just So Stories, 9, 11-2, 16, 49, 142, 192

Kamin, Leon, 126
Kano, Takayoshi, 69, 217
Kant, Immanuel
 philosophy of, 141-2, 148, 150, 181-2, 184, 187, 218
 race of devils (group size problem), 182, 187
Kasparov, Garry, 185
Kitcher, Philip, 9, 143-4, 146, 199, 200, 201, 218
Kurzweil, Ray, 185
Kymlicka, Will, 153, 218

Index

Lamarck, Jean-Baptiste, 3
Lamarckism, 21
Langur (monkey), xii, 64-5
Lewontin, Richard, 9, 83, 87, 89, 99-100, 199, 218
Linnaeus, Carolus, 51, 208
Living Without Free Will, 158, 219
Lombroso, Cesare, 99-100
Lowell, Percival, 177
Lumsden, Charles, 38, 218
Lust, 129

Malingering brute hypothesis. *See* Owen, Richard
Man's Place in Nature, 24, 73, 197, 217
Mandeville, Bernard de, 92
Marxism, 96, 109
Materialism, 22, 138, 140, 146, 157, 193, 198, 208
Maynard Smith, John, xiii, 9, 18, 19, 20, 24-5, 27-8, 31-2, 38, 44, 49, 59, 74-5, 77, 80, 82-4, 86-8, 92-3, 114, 116, 131, 140-2, 175, 177, 181-2, 200, 202, 208, 218
Mayr, Ernst, 87
Medawar, Peter, 30, 125-6, 131, 218
Memes, 29-31, 74, 78, 166-7, 175, 188
Mendel, Gregor, 87
Midgley, Mary, 91, 218
Midianites, 167-8, 170, 175, 203
Milton, Katharine, 69
Mineau, Pierre, 62, 72, 219
Mole-rat
  Damaraland mole-rat, 44
  naked mole-rat, 43-4, 219
Moore, James, 23, 138, 157, 169, 215
Moral Animal, The, 11, 112, 198, 222

Moravec, Hans, 185
Morph, 43, 135, 219
Moses, 130, 167-71, 203
Mozu, 48
Multilevel selection, 42, 60, 82-5, 87-90, 94, 99-100, 193, 199, 201, 208

Natural selection
  adaptation, 4, 6-7, 12, 25-6, 31, 50, 60, 78, 83, 130, 135, 193
  genic selection, xi, xii, 2, 8-10, 16, 18-9, 22, 25-7, 29, 31-6, 40, 42, 44-50, 55, 60, 62, 66, 68, 70-3, 82-94, 97, 110, 112, 130, 165, 170, 178-9, 181, 189, 192, 195, 199, 201, 203, 209, 215
  group selection, 17, 20, 25, 27, 42, 45-6, 48, 51, 60, 83-4, 87, 192
  individual selection, 17-20, 45
  sexual selection, 207
Nature, 210
Newton, Isaac, 176
Nietzsche, Friedrich, 2, 37, 94, 137, 150, 154, 157, 166, 169, 171, 173-4, 182, 184, 204, 205, 219
Noble Lie (Plato), 80, 154
Nonshared environment, 198
Nuffield Council on Bioethics, 210

O'Hear, Anthony, 193, 219
Old Testament, The, 167-9, 171
Orang-utan, 39, 70
Origin of Species, 3, 17, 21, 23, 37, 140, 214, 217, 221
Origins of Virtue, The, 13, 16, 220
Owen, Richard, 134-5, 197

Paley, William, 4

227

Index

Palmer, Craig, 135, 220
Pan paniscus. See Bonobo
Pan troglodytes. See Chimpanzee
Paradis, James, 62, 65, 219
Parish, Amy, 71
Pascal, Blaise, 75, 158-62, 179, 219
Pascal's Wager, 159-62
Pereboom, Derk, 158, 219
Perfect pitch, 132
Petrinovich, Lewis, 50-1, 219
Phenotype, 50, 77, 124, 174, 215
Phenotypic evolution, 50
Phenylketonuria (PKU), 132
Phylogenetic constraint, 52
Pinker, Steven, 9, 26, 49, 94, 97, 103, 120-1, 146, 152-3, 157, 200, 208, 219
Plato, xiii, 1, 39, 80, 98, 146-7, 149, 154, 183, 202, 219
Plinian Society, 157
Ploidy
　diploid, 14, 18, 32, 43, 49, 79, 165
　haplodiploid, 14-6, 19, 43, 165
　haploid, 14, 43
Pollack, Robert, 105, 219
Popper, Karl, 97, 110, 140-1, 207, 221
Prisoner's Dilemma, 27, 179, 182-3
Punctuated equilibrium, 83

Radford, Tim, 96, 219
Rand, Ayn, 74, 110
Rape, 52, 61-2, 70-2, 81, 128, 135, 166, 169, 171, 202, 219, 220
Reason, 179–84
Richards, Robert, 20, 182, 219
Richerson, Peter, 28, 214
Ridley, Mark, 5, 7, 13, 34, 77, 79, 114
Ridley, Matt, 13, 34, 49, 78, 87, 97, 137, 144, 200
Robertson, Ian, 123, 131, 220
Robinson, Daniel, 119, 122, 124, 154, 156-7, 220
Rose, Steven, 89, 215
Ruse, Michael, 7, 19

Sagan, Carl, 177, 220
Saltation, 51, 57
Sartre, Jean-Paul, 193
Schizophrenia, 5, 103, 125
Sea lion, New Zealand, 67
Seal, elephant, 64, 67, 186
Seal, grey, 67
Segerstrale, Ullica, 87-8, 220
Selfish gene theory. *See* Natural selection, genic selection
Selfish Gene, The, 5, 8, 13, 29-31, 38, 45, 47, 87, 89, 91-2, 100, 165, 177, 183, 188-9, 215
Sexuality
　ambisexuality, 127
　bisexuality, 127
　homosexuality, 6-7, 62, 102, 125, 127-9
Shark, sand tiger, 66
Shaw, George Bernard, 98
Sibley, Charles, 53, 220
Siege mentality within gene-selectionism, 90
Singer, Peter, 75, 92-4, 180-1, 220
Snake, 52
Sober, Elliott, 42, 45, 220
Social insects, 6-7, 14-6, 33, 42-5, 148, 172, 179, 185-6, 199-200, 203
Sociobiology. *See* Human sociobiology
Socrates, 149, 171, 208
Son of sociobiology, 9, 131

Index

Sparta, 98
Squirrel, 65
Stanford, Craig, 69
Stepchildren, abuse of, 113-8, 135
Subversion from within, 32, 46-7, 84
Symons, Donald, 11, 57, 220
Sympathy
   capacity for simulation, 188-91, 194
   empathy, 116-8, 166, 188, 190-1, 193
   Sartre and the gaze, 193
Szathmary, Eors, 28, 75, 183, 218

't Hooft, Gerard, 143, 220
Thatcher, Margaret, 74
Thornhill, Randy, 135, 220
Tooby, John, 11, 213, 220
Transmutation, 3, 213
Trigg, Roger, 92-3, 221
Trivers, Robert, 19
Turner, John, 100, 171, 221
Twinsburg, Ohio, 107

U.S. Department of Justice, 109
Urmson, J. O., 2, 221
Utilitarianism, 153

van Inwagen, Peter, 147, 221
van Noordwijk, Maria, 70, 221
van Schaik, Carel, 70, 221
Voltaire, 160-1

Wallace, Alfred Russel, xii, xiii, 3, 19, 21-4, 31, 48, 79-80, 104, 107, 172, 179, 201, 208, 211, 214, 221
Walleye, 67
Wamba National Park, 69, 71
Wells, H. G., 177

Wiener, Norbert, 142
Wilberforce, Samuel, 139, 155, 201, 221
Williams, George C., xii, xiii, 8, 12, 15, 18, 20, 24-8, 30-1, 34, 41-2, 46, 51-2, 59-61, 63-5, 68, 74, 76-80, 82-7, 92-3, 127, 155, 181, 187, 196, 198, 200-1, 208, 210, 212, 219, 221
Wilson, David Sloan, 42, 45, 220
Wilson, Edward O., 5-9, 12, 39, 59, 75, 81, 89, 138, 142, 144, 172, 184, 200, 205, 221, 222
Wilson, Margo, 113, 134-5
Windsor, Elizabeth, 73-4
Winston, Robert, 108
Wojtyla, Karol, 73-4
Wolf, Susan, 148, 149, 153, 157
Wright, Lawrence, 104, 222
Wright, Robert, 11, 34, 90, 111, 192, 198, 222
Wright, Sewall, 25
Wynne-Edwards, V.C., 26-7

229